Sequence Analysis in Molecular Biology

Treasure Trove or Trivial Pursuit

Sequence Analysis in Molecular Biology

Treasure Trove or Trivial Pursuit

Gunnar von Heijne
Research Group for Theoretical Biophysics
Department of Theoretical Physics
Royal Institute of Technology
Stockholm, Sweden

ACADEMIC PRESS, INC
Harcourt Brace Jovanovich, Publishers
San Diego New York Berkeley Boston
London Sydney Tokyo Toronto

ACADEMIC PRESS, INC.
1250 Sixth Avenue, San Diego, California 92101

United Kingdom Edition published by
ACADEMIC PRESS INC. (LONDON) LTD.
24–28 Oval Road, London NW1 7DX

Library of Congress Cataloging in Publication Data

Heijne, Gunnar von.
Sequence analysis in molecular biology.

Bibliography: p.
Includes index.
1. Amino acid sequence—Data processing. 2. Nucleotide sequence—Data processing. I. Title.
QP551.H43 1987 574.87'328 87-1292
ISBN 0–12–725130–8 (alk. paper)

PRINTED IN THE UNITED STATES OF AMERICA

88 89 90 9 8 7 6 5 4 3 2

To Anna

And to Nisse and Erika
for making sure I didn't waste too much time on sleep

Contents

Preface

Gimme an A . . . **A!**
Gimme a C . . . **C!**
Gimme a G . . . **G!**
Gimme a T . . . **T!**
What's that spell? . . . . **Grants, Grants, Grants!!**

From "100 Ways Not to Open a Molecular Biology Meeting"
Biopress, 1987

Sequencing is no longer an art; it is graduate student routine. From humble beginnings (the 30-residue B-chain of insulin) to today's estimated 2 Mbp per year (Mbp is jargon for mega base pairs, i.e., 10^6 base pairs), the number of available protein and DNA sequences has grown exponentially. At present, about 10 Mbp's worth of sequence information is stored in the main data banks, information that could be tapped much more efficiently and with much more imagination than it is today.

This book deals with sequence analysis on the computer. One of its aims is to serve as a brief survey of what one can do with protein and DNA sequences either directly on a microcomputer or by using one of the main sequence/programs data banks such as BioNet or the Wisconsin package. Equally important, the book traces the origins of some of the ideas that have come to be embodied in these programs from both biological and methodological points of view: What do the standard sequence analysis algorithms really analyze, and to what degree can we trust their outputs?

It is my hope that a short book on methods for sequence analysis of both DNA and proteins will be more useful to the reader than an in-depth review of a particular subfield and that it will help fill a small gap in the available literature. Personally I prefer a short monograph by one or two authors to a massive compendium of poorly related and often redundant multiauthor reviews.

I have written this book for the molecular biologist who often uses methods and programs without really knowing what is going on inside his electronic black box rather than for the relatively few who spend a good deal of time crouching and (sometimes) swearing over the computer. For this reason, I have tried throughout to concentrate on the main ideas and the reliability of the methods rather than on details of computation.

I have also tried to present sequence analysis as something of an intellectual challenge in the hope that a few more molecular biologists will be encouraged to put their gels aside now and then and spend some time thinking—with or without computer assistance—about what their sequences really seem to be telling them, and how they should go about making the messages more audible.

Thanks are due to Dr. Lars Abrahamsén for ploughing through early versions of the manuscript; to Dr. Christian Burks for Appendix 1; to the Aspen Center for Physics, where some of the work was done; and to the Swedish Natural Sciences Research Council for grants.

Gunnar von Heijne

Sequence Analysis in Molecular Biology

Treasure Trove or Trivial Pursuit

Chapter 1

Introduction: The Abnormalin Story

All the gels have been read. The complete 6.021-bp sequence has been duly put together on your microcomputer by one of the postdoctoral students. Only a couple of loose ends are left. You can already see the paper: "Cloning and Sequencing of the Gene for Human Abnormalin—a Novel Class of Neuropeptides."

Oh yes, you had better get one of the students to run some of the standard programs: codon usage, a full-scale homology search (who knows, an oncogene may pop up . . .), a hydrophobicity (or was it hydrophilicity?) analysis, maybe a Chou–Fasman prediction. These days, you seem to need a couple of impressive computer-generated figures to make a paper on sequencing more palatable. . . .

Abnormalin, as is well known, was discovered accidentally in 1958 when various preparations of human pancreatic homogenates were tested for their possible use as a bleaching agent in commercial detergents ("Gives Your Linen the *Natural* Whiteness It Deserves")—one of the lab technicians was careless with a pipette and immediately took off into hyperspace. The active substance was later found to be concentrated in brain preparations, was purified to homogeneity in the early 1970s, and a partial amino acid sequence was published in 1981.

It was not until DNA technology was brought into play, however, that a complete sequence could be assembled. In your lab, a new microcomputer, complete with all the latest molecular biology software and nucleic acid and protein sequence data banks, had just been installed; abnormalin was to be its first test. So, manual in one hand (or resting heavily on a nearby table, to be precise) and the partial amino acid sequence in the other, you sat down for your first session: designing a suitable oligonucleotide probe for fishing out the abnormalin gene. This was accomplished in no time by the PROBE program, which gave you suggestions for probes of minimal ambiguity biased toward the known codon preferences of human genes.

With this probe you (well, not you exactly, but somebody in the lab) then screened a human gene library, picked out three positive clones, accomplished restriction site mapping and dideoxy sequencing in no time (thanks in no small measure to the SEQME program), pinned down the most likely coding regions and splice sites (programs YESCODE and SPLCSTS), checked for promoters, terminators, poly(A) sites, and putative initiator AUGs (PROM, TERM, PADD, and (KOZAK), calculated the most stable mRNA secondary structure (HAIRPINS), detected a secretory signal peptide and a membrane-spanning region in the deduced protein chain (D-TO-P, SECRET, and AMPHIPATHETIC), predicted three strong antigenic sites (ANTI), a possible glycosylation site (SWEET), and three α-helices (OL'FAITHFUL) in the extracellular domain, and finally detected a highly significant homology ($P < 10^{-8}$) to a recently published oncogene sequence (FINDONC). Success and world-wide recognition is obviously imminent—except that your \$500 printer refuses to have anything to do with the manuscript that your state-of-the-art word processor is trying to push down its throat. . . .

The abnormalin story touches upon a few of a vast number of sequence analysis algorithms that have been developed over the past decade. Some of these are now included in almost any commercially available software package; others are to be found only in the specialized literature or in not very user-friendly versions obtainable directly from the authors.

This book is an attempt to give an overview of what has been accomplished so far in this area, and to describe the methods that have been developed. It starts with a description of the main nucleic acid and protein sequence data banks (Chapter 2) and a short section on the "housekeeping aids" that the computer can provide during a sequencing project (Chapter 3). Chapters 4 and 5 deal with nucleic acid and protein sequence analysis, and make up the bulk of the text. Chapter 6 treats algorithms for homology searching and sequence alignments. Chapter 7 presents some selected examples of how computer modeling can help decide whether an observed sequence pattern is significant or not, and how computer simulation is sometimes used to get a feeling for the behavior of intrinsically complex sequence-dependent processes. Chapter 8, finally, contains some comments on the role of theoretical sequence analysis in molecular biology.

Chapter 2

The Collector's Dream: From Dayhoff to Data Banks

"Biology is mere stamp-collecting," is a favorite quote among physicists and chemists. And indeed, if the heroic age of natural history and early Darwinian biology was preoccupied with filling museums and schools with box upon box of dead specimens, modern molecular biology has a similar predilection for producing giant computer data banks with nothing but As, Ts, Gs, and Cs in them. If the biological stamps of yesteryear were somewhat dusty, though at least appealing to the imagination, those of today have been reduced to an epitome of scientific dryness.

Nevertheless, sequence data banks embody vast amounts of biological information. In principle, the linear script of DNA and protein sequences that is so tiring to the eye tells all there is to know about the genetic makeup of an organism — its genetic program — and also contains vestiges of the evolutionary history of the species. Of primary importance, then, is to make this store of molecular information as accessible and as easy to analyze as possible.

It all started in 1951. Through a combination of limited proteolytic digestions and chemical analysis of the resulting peptides, Sanger and Tuppy (1951) were able to derive the complete primary sequence of the B-chain of bovine insulin, a short polypeptide of 30 residues. Protein sequencing had come of age, but it was slow and tedious work and the number of sequences increased only at a low rate.

In 1965 Holley *et al.* (1965) derived the complete nucleotide sequence of an alanine tRNA from yeast. Again, a lot of chemical ingenuity and long hours in the lab were required to determine the sequence of even the shortest stretch of RNA. It was not until the mid-1970s that the real methodological breakthrough came with the first modern DNA sequencing techniques (Maxam and Gilbert, 1977; Sanger *et al.*, 1977); from now on the molecular revolution in biology was to gain an ever-increasing momentum.

At first, the volume of sequence data could be easily managed manually and collected in standard book format—most biological libraries still have one or other edition of Dayhoff's *Atlas of Protein Sequence and Structure* (Dayhoff, 1972) collecting dust on a shelf. With the advent of rapid DNA sequencing, however, it was soon obvious that sequence data could no longer be handled in printed form. Thus, in the early 1980s work was initiated to collect and distribute computerized data banks of protein and nucleic acid sequences: the EMBL Data Library at the European Molecular Biology Laboratory in Heidelberg, F.R.G.; the GenBank* Genetic Sequence Data Bank at Los Alamos National Laboratory in New Mexico, U.S.A.; and the National Biomedical Research Foundation Protein Sequence Data Base (NBRF) at Georgetown University, Washington D.C., U.S., to mention the most important ones.

Since their inception, these libraries have multiplied in size many times over (Fig. 2.1). Current estimates put the number of bases sequenced per year at around two million (Burks *et al.*, 1985). The April 1986 release 8.0 of the EMBL library counts more than 6.3 Mbp with a doubling time of 2 years. This release cites more than 4400 references and has 6400 different entries. Approximately 50% of the sequences are of eukaryotic origin, with viral and phage sequences accounting for another 20%. Some 10% of the *Escherichia coli* genome is already in the data bank, as is 1% of the *Saccharomyces cerevisiae* genome. Sequences of human origin, although being the largest group in the data base, still represent less than 0.1% of the human genome. Roughly 50% of the sequence data is from known protein-coding regions; introns contribute 12% (Bilofsky *et al.*, 1986). In Release 9.0 (September 1986), the data base has grown to almost 8 Mbp—the 10 Mbp mark is thus less than a year away, and has probably been reached by the time you are reading this.

A typical GenBank entry is shown in Fig. 2.2. The LOCUS linetype gives the name of the entry, the number of bases, and the date of entry

*Trademark of the U.S. Department of Health and Human Services.

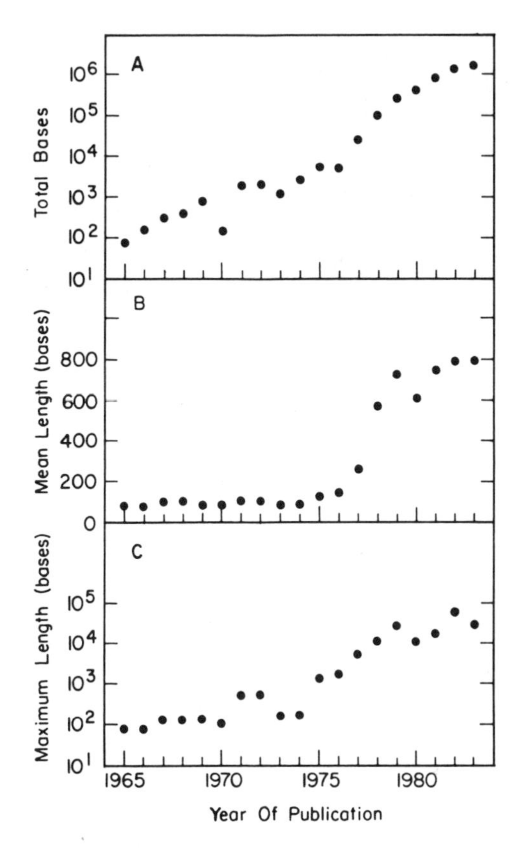

Fig. 2.1. The growth of nucleic acid sequence data. (A) Total bases published *versus* year of publication. (B) Mean published sequence length. (C) Maximum published sequence length. (From Burks *et al.*, 1985.)

or update. The ACCESSION number, although not as informative as the LOCUS name, is a unique identifier that is never changed (the LOCUS name *is* sometimes changed), and it is the same in both the GenBank and EMBL data banks. It is thus recommended that references to a particular entry include this number.

Other important linetypes are FEATURES and SITES, which hold information on coding regions, start and stop signals for transcription and translation, conflicting reports on the sequence, known mutations, etc. EMBL and NBRF entries are similarly organized. GenBank and EMBL regularly exchange data, and the two are almost identical in content.

```
LOCUS       HUMOPS        6953 bp ds-DNA            entered    08/01/85
DEFINITION  Human opsin gene, complete cds.
ACCESSION   K02281
KEYWORDS    opsin; rhodopsin.
SOURCE      Human (individual J.N.) germline DNA, clone roJHN.
ORGANISM    Homo sapiens
            Eukaryota; Metazoa; Chordata; Vertebrata; Tetrapoda; Mammalia;
            Eutheria; Primates.
REFERENCE   1  (bases 1 to 6953; exons and flanking sequence only)
AUTHORS     Nathans,J. and Hogness,D.S.
TITLE       Isolation and nucleotide sequence of the gene encoding human
            rhodopsin
JOURNAL     Proc Nat Acad Sci USA 81, 4851-4855 (1984)
REFERENCE   2  (bases 1 to 6953)
AUTHORS     Nathans,J. and Hogness,D.S.
JOURNAL     Unpublished (1985) Dept of Biochemistry, Stanford U., Stanford CA
COMMENT     The five coding regions of the human opsin gene are 89.7%
            homologous to the five coding regions of the bovine opsin gene.
            The 5' untranslated region is 77% homologous to the bovine
            sequence.  The four introns present in both genes occur at
            precisely analogous positions and are of comparable lengths.
            Potential CAAT and TATA boxes are found at positions 122-127 and
            171-177 respectively.  Potential polyadenylation signals are found
            at positions 5642-5647 and 6698-6903.
FEATURES       from  to/span     description
    pept        295      655     opsin, exon 1
               2439     2607     opsin, exon 2
               3813     3978     opsin, exon 3
               4095     4334     opsin, exon 4
               5168     5278     opsin, exon 5
SITES
    ->mRNA      200        1     opsin mRNA exon 1 start site 1 [1]
    ->mRNA      202        1     opsin mRNA exon 1 start site 2 [1]
    ->pept      295        1     opsin cds start
    pept/IVS    656        0     opsin cds exon 1 end/intron A start
    IVS/pept   2439        0     opsin cds intron A end/exon 2 start
    pept/IVS   2608        0     opsin cds exon 2 end/intron B start
    IVS/pept   3813        0     opsin cds intron B end/exon 3 start
    pept/IVS   3979        0     opsin cds exon 3 end/intron C start
    IVS/pept   4095        0     opsin cds intron C end/exon 4 start
    pept/IVS   4335        0     opsin cds exon 4 end/intron D start
    IVS/pept   5168        0     opsin cds intron D end/exon 5 start
    pept<-     5278        1     opsin cds end
BASE COUNT     1524 a    2022 c    1796 g    1611 t
ORIGIN      1 bp upstream of BamHI site.

    PRIMATE:HUMOPS  Length: 6953  29-OCT-1986 12:11  Check: 3809  ..

       1  GGATCCTGAG TACCTCTCCT CCCTGACCTC AGGCTTCCTC CTAGTGTCAC

      51  CTTGGCCCCT CTTAGAAGCC AATTAGGCCC TCAGTTTCTG CAGCGGGGAT
                                ..........

    6901  CCAACTTTGG GGTCATAGAG GCACAGGTAA CCCATAAAAC TGCAAACAAG

    6951  CTT
```

Fig. 2.2. A typical GenBank entry. Only the first and last lines of the sequence are shown.

Both GenBank and EMBL provide four annual releases, with limited updates in between. They are distributed on standard magnetic tape and on floppy disks; GenBank is also available on-line. Printed compendia of both data banks are published periodically, the latest one in 1985 (Armstrong *et al.,* 1985).

The NBRF Protein Sequence Data Base is a computerized descendant of the Dayhoff Atlas. It contains sequences obtained by protein as well as DNA sequencing, and is thus an important source of information regarding posttranslational modifications of proteins. It is also much easier, faster, and less costly to search for protein sequences directly in the NBRF data bank, rather than via the cumbersome detour through one of the DNA sequence libraries. The Protein Identification Resource (PIR) maintained and distributed by the National Biomedical Research Foundation has developed a large number of commands (the PSQ program) for searching the data base for specific proteins, homologous sequences, particular classes of proteins, etc. (George *et al.,* 1986; Foley *et al.,* 1986; Orcutt *et al.,* 1983).

The Institut Pasteur in Paris also provides a protein sequence data bank called PGtrans that is generated by automatic translation of GenBank (Claverie and Sauvaget, 1985).

A major problem plaguing the general-purpose data banks described so far is their inability to keep up with the amount of sequence data produced. GenBank, which seems harder hit than EMBL, has a backlog of published sequences yet to be entered going back 1–2 years (Lewin, 1986). With an estimated doubling time of 2 years, this means that there are almost as many published sequences not in the databases as there are in them. This is shown graphically in Fig. 2.3, where the number of literature references in NBRF Release 7.0 (November 1985) is plotted *versus* year of publication; clearly, there is an embarrassing lack of data from more recent years.

One obvious way to speed up the process of data entry is to have authors submit their data directly to the data bank in conjunction with the publication process proper. Database administrators have time and again asked for either clean hard copy data or, even better, for data submitted in computer-readable form on floppy disks or via teletransmission—to no great avail, as it seems, since the response to such pleas has been poor at best (Lewin, 1986).

Another important but often overlooked responsibility of authors is to inform the data bank of any errors in the original sequences; such errors are often "merely noted within the group that published the

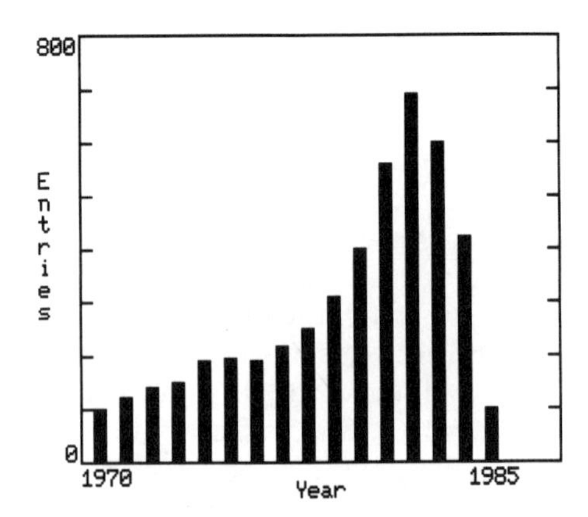

Fig. 2.3. Number of entries in the NBRF Protein Sequence Data Bank (Release 7.0 + update 14.0, December 1985) *versus* year of publication. Note the backlog in later years.

original sequence and passed by word-of-mouth to other groups interested in that sequence" (Burks *et al.,* 1985).

Problems such as these have prompted individual investigators or groups to collect and maintain sequence data bases of a more specialized character. Recently, the Committee on Data for Science and Technology under the International Council of Scientific Unions (CODATA for short) has formed a Task Group on Coordination of Protein Sequence Data Banks which, among other things, will maintain a directory of such data bases (Lesk, 1985; see also Appendix 1).

In conclusion, the enormous growth of sequence data over the past decade has created problems of data handling as well as interesting opportunities for conducting biological research on the computer. Some of the tools available for this will be described in the following pages.

Chapter 3

Sequencing on the Computer: Programs That Make Lab Work Easier, Faster, and More Fun (?)

Although somewhat outside the main theme of this book, a short chapter on how computers can be used to aid in sequencing work seems unavoidable, given the large number of programs available and the fact that many a molecular biologist has had his first serious encounter with the world of electronic wizardry in the context of managing a sequencing project. And after all, who can resist gel readers that beep when you make a mistake, or speech synthesizers that entertain the baffled listener with endless runs of flat-spoken As, Gs, Cs, and Ts?

I. DESIGNING OLIGONUCLEOTIDE PROBES

Oligonucleotide probes are used increasingly to pick up genes or cDNA clones from genomic or cDNA libraries when the (partial) amino acid sequence is known (Itakura *et al.,* 1984). Such probes can also be used as primers which, when annealed to complementary mRNAs, can be selectively extended into cDNAs (Houghton *et al.,* 1980). However, due to the degeneracy of the genetic code, reverse translation of the amino acid sequence into a corresponding DNA sequence does not yield a unique sequence; rather, a whole set of possible coding sequences corresponds to any given amino acid sequence.

The problem facing the experimenter is, thus, to find a probe or mixture of probes that maximizes his chances of successful hybridization, and at the same time minimizes the amount of time and money that has to be spent on making the probes.

Two approaches have been suggested. The "cocktail" strategy amounts to selecting a segment of the amino acid sequence that corresponds to a minimal set of probe sequences. An appropriate mixture of these probes, one of which will be perfectly homologous to the sequence sought for, is then used for hybridization. The "one-shot" strategy, on the other hand, relies on using only one unique probe, constructed with the aid of known DNA sequence regularities such as codon preferences and GC avoidance. This probe, if long enough, should hybridize preferentially with the target sequence. Obviously, in both strategies, computer analysis can greatly facilitate the search for optimal probe sequences.

In its simplest version, the search procedure is limited to finding a set of probes of given lengths with the least possible degeneracy simply by scanning the amino acid sequence and noting the number of alternative codons in the corresponding oligonucleotide as one moves along the chain (Lewis, 1986). Codon usage statistics can also be included, thus attaching a probability-of-occurrence value to each probe (Raupach, 1984).

A somewhat more advanced algorithm allows the user to specify the way in which he plans to synthesize the probes: by adding one monomer or a mixture of monomers at each step; by adding monomers or (mixtures) of dinucleotides; or by adding monomers, di- or trinucleotides. This makes it possible to find a probe mixture that minimizes the amount of work spent on its synthesis (Yang *et al.,* 1984). It is also easy to add a rough estimate of the dissociation temperature of each probe to programs such as these (Yang *et al.,* 1984).

In some instances, nonstandard bases such as deoxyinosine can be incorporated opposite to a degenerate site in the hope that they will be less discriminating than the standard ones. Thus, inosine apparently can be used to help suppress A/C and G/T ambiguities (Martin and Castro, 1985). This possibility can also be incorporated into probe-design programs (see Yang *et al.,* 1984, which has an option for using G at U/C and T at G/A ambiguities).

A true optimization of the probe in terms not only of degeneracy but in terms of length, codon usage, GC avoidance, and expected signal-to-noise ratio (hybridization to target over background) is a fairly complex problem, however (Lathe, 1985), and does not seem to have been automated so far.

In this context it is interesting to note that the DNA sequence data banks have now grown to such a size that most any oligonucleotide of length less or equal to seven can be found sandwiched between suitable restriction sites. It is thus possible that gene libraries rather than chemical synthesis can occasionally be used as a source of oligonucleotides in the future (Tung and Burks, 1986).

II. CONSTRUCTING RESTRICTION MAPS

The construction of restriction maps from single and double digests quickly becomes unmanageable without computer assistance. The number of possible single-digest maps grows as the factorial of the number of fragments in the data, and the combinatorics become even worse when one takes the errors in the fragment-size determination into account.

Many integrated packages have a routine for entering the positions of the fragments directly from the gels using a digitizer, whereupon their sizes are determined by interpolating on a nonlinear curve fitted to the known standards (e.g., Duggleby *et al.,* 1981). At this step, one can also obtain an estimate of the likely errors associated with the calculated sizes.

The actual construction of the map is straightforward in principle: Generate all possible single-digest maps and check all combinations of these maps against the double-digest data. In practice, this approach rapidly outgrows the computing power even of large computers, and one has to devise algorithms that rapidly sift through partial maps to screen out those that do not fit the data. As an example, one program for constructing linear maps (Pearson, 1982) first considers all possible maps for the two single digests with the smallest number of fragments. All possible double digests from these maps are then evaluated against the data, a measure of the goodness-of-fit is calculated for each, and only those that have a calculated fit above some given value are retained. The solutions remaining after this step are then used together with all possible maps from the third single digest, etc., until at the end only one or a few maps remain. With some modifications, a similar procedure can also be used for circular maps.

In another approach (Durand and Bregegere, 1984), all the linear single-digest maps are constructed in parallel by tentatively adding fragments (starting with the shortest ones) and checking against the double digests until an acceptable one is found (given a certain leeway for errors in fragment size). A routine such as this requires a certain amount of backtracking since a fragment that is added to a partial map at one step may later prove not to be compatible with the double digests, but the method apparently is very efficient and can be used even when the number of fragments is very large.

A final example is provided by the algorithm of Nolan *et al.* (1984), which starts by finding all possible groupings of the double-digest fragments from a pair of restriction enzymes that add up (within a given error) to fragment sizes found in the single digests. These groups are then checked against the data to weed out the incompatible ones according to a set of logical rules. According to the authors, this method finds all maps compatible with the data for a pair of enzymes in a matter of a few minutes on a microcomputer.

III. SEQUENCING DNA

Once an appropriate restriction map has been worked out, additional advice on how best to proceed with the actual sequencing can be obtained from the computer. Programs have been developed that derive a sequencing strategy that minimizes the number of gels that have to be run and maximizes the number of new bases read from each gel, given the maximum number of bases that the lab can currently read from a gel and the minimum size difference between fragments that can be separated cleanly (Bach *et al.,* 1982).

In a typical run, you are asked to provide the restriction map and define the region you want to sequence. The program then analyzes all possible separable double-digest fragments, and ranks them according to the number of new residues that can be read from each. This process is continued until the whole region of interest has been sequenced, or when no more fragments can be obtained. As sequencing proceeds, new restriction sites found in the parts already sequenced can be incorporated to optimize the remaining steps.

Many of the integrated software packages for DNA data management include digitizers for direct entry of nucleotides from the gel. Some kind of audible signal or confirmation through a speech synthesizer is often used to verify the gel reading or to alert the user to discrepancies between the first and second entry of a given gel. The digitizer can also be used to enter data from restriction fragment gels.

Once sequencing is under way, the main problem is data handling. New gels have to be added to those already stored in the data base, unwanted sequences such as those originating from the cloning vector must be screened out, and any overlaps between the new gel and previously sequenced ones must be noted. Finally, overlapping gels must be combined into "contigs" or "melds," and an appropriate record of uncertainties in the original gel readings, or mismatches and insertions created during the construction of consensus sequences in overlapping regions, must be kept.

This whole sequence of operations can be performed automatically by a number of programs (Staden, 1982b, 1986; Grymes *et al.,* 1986). The most critical part of programs of this kind is the algorithm used for finding acceptable overlaps: An erroneous meld will lead to serious problems, whereas a failure to detect a true overlap will result in unnecessary extra work.

As an example of a recent meld algorithm, the GEL program described by Grymes *et al.* (1986; part of the IntelliGenetics software, see Appendix 2) searches for matches that (i) start within the first 10 bases of either sequence being compared, (ii) start with an exact dinucleotide homology, (iii) can be extended only allowing for a given number of insertions and mismatches followed by a dinucleotide homology, and (iv) are terminated when a mismatch is not followed by (at least) a dinucleotide homology. For any given match the percent similarity is calculated, and the match is rejected if this value is below a user-definable limit (which can be made to depend on the length of the match). If more than one acceptable match between two sequences is found, the best (highest-scoring) one is chosen for the alignment.

In this way, individual gels can be combined into melds, two melds can be merged into a larger meld when a joining gel is found, and finally a single meld containing the consensus sequence for the whole project will result. Programs such as these, although originally written for relatively large mainframe computers, are now available for microcomputers of the IBM PC XT/AT class.

IV. FINDING RESTRICTION SITES

Searching for restriction sites in long segments of DNA is an obvious task best entrusted to a computer. The principle behind such a search is simple enough: to find strings of nucleotides that match the recognition sequence of a particular restriction enzyme. Depending on the algorithm, the search can be performed more or less rapidly, but details of programming need not concern us here.

Most programs of this kind come with a predefined list of restriction enzyme cut sites, most often from the compilation by Roberts (1985). This list can be altered to suit the stock presently on the lab shelf, and new entries can be added. The output from a restriction site search can also be more or less elaborate: linear (Fig. 3.1a), circular (Fig. 3.1b), or simply a listing of the sites found. Various simple but tedious calculations such as determination of fragment sizes from multiple digests, or simulations of gel separations with identification of labeled, unlabeled, and semilabeled fragments can also be performed by most of the commercially available programs listed in Appendix 2.

An interesting use of restriction site analysis is to find sites in a given protein-coding sequence that can be mutated to introduce a new restriction site without changing the amino acid sequence. A program performing such a search has been described, and this program can also be used to find out which amino acid replacements can be made in a given position with a simultaneous introduction of a new restriction site (Arentzen and Ripka, 1984). One can also do the converse, i.e., design DNA segments coding for a given amino acid sequence that *lack* particular restriction sites (van den Berg and Osinga, 1986).

V. TRANSLATING DNA INTO PROTEIN

DNA-to-protein translation in any or all possible reading frames is a function offered by all available software packages, though not all include the possibility to use nonstandard as well as the normal genetic codes. In systems that include one or other of the main data banks it is often possible to get automatic translation of given DNA segments using the information on coding regions and splice sites from the FEATURES linetype (Chapter 2).

Simple calculations on the resulting amino acid sequence are often available: molecular weight; net electric charge as function of pH;

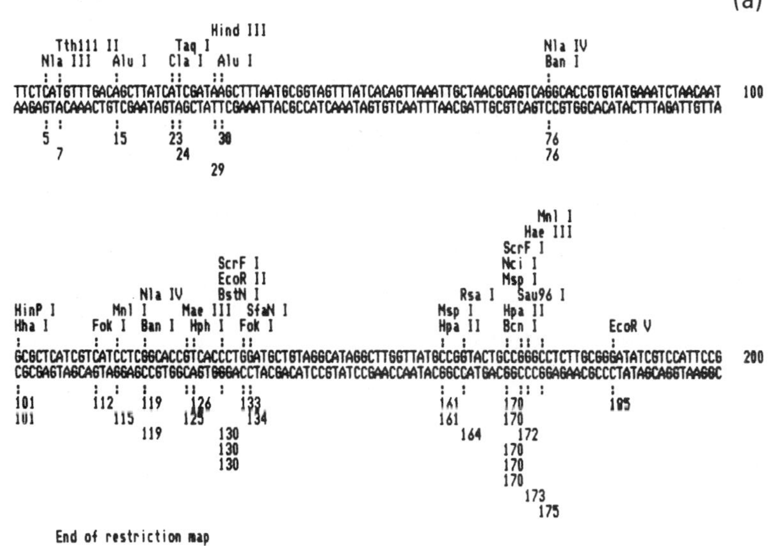

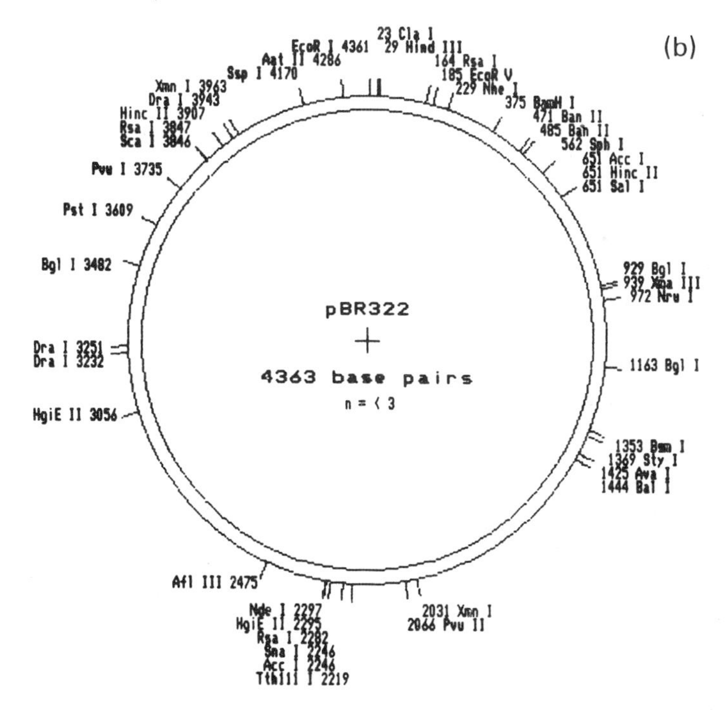

Fig. 3.1. Linear (a) and circular (b) restriction maps produced by the DPSA program of Marck (1986).

titration curves; simulations of proteolytic digestions with display of fragment lengths and molecular weights; isoelectric points; and predicted HPLC retention times.

VI. SIMULATING CLONING EXPERIMENTS

Some of the more advanced systems [e.g., IntelliGenetics CLONER program and the CAGE/GEM program (Douthart *et al.,* 1986)] can perform veritable simulations of cloning experiments on the computer. Here, the region of interest is displayed graphically, with overlays indicating genetically relevant features such as coding regions, restriction sites, promoters, and symmetrical sequences. By zooming in and out, one can go from the global picture to the details of the sequence and back again.

Cloning simulations are performed by cutting with restriction enzymes, separating fragments, filling in overhangs in various ways to create, e.g., blunt ends, inserting new fragments into appropriate restriction sites with a check for compatibility of the ends before ligation, etc. As the simulation progresses, the genetic information in the overlays is automatically updated to indicate what changes have been made. These programs require a fairly large computer and advanced graphics displays.

In summary, much tedious data handling can now be performed quickly, accurately, and relatively cheaply on a computer. A number of integrated software packages, sometimes complete with appropriate hardware, are offered commercially (see Appendix 2). Many programs more limited in scope can also be obtained directly from the scientist–programmer, often for only a nominal fee; such programs are reviewed in, e.g., Jungck and Friedman (1984), Korn and Queen (1984), and Mount (1985).

Chapter 4

Nucleotide Sequences: What You Can Do With Your Sequence Once You Have It

From the point of view of "doing molecular biology on the computer," the real fun starts only when all the "trivialities" of the bench have been successfully struggled through and the DNA sequence flashes across the monitor in all its literary beauty. And, unfortunately, this is where the expertise of the experimenter often meets its limits. It is all too easy to run the whole gamut of sequence-analysis programs available in the molecular biology software package that the lab just bought and be impressed with the apparent finality about the messages from the computer: List of promoters found: . . . , List of terminators found: . . . , List of coding regions:. . . . The advanced graphics displays, all added with the best intentions of making the program "user-friendly and easy to grasp," make it even harder to doubt the computer's verdict.

And yet, if there is one take-home message from this book, it is DON'T EXPECT YOUR COMPUTER TO TELL YOU THE TRUTH. Used with care and with an appreciation of how well or badly the particular algorithm you are running is expected to do on the kind of questions you are trying to have it answer, computer analysis of your sequence can help a lot and is indeed unavoidable in many situations. But such methods are also dangerous in that they might fool you into publishing totally worthless results when they are applied improperly —as anyone who has ever helplessly witnessed his or her prediction algorithm being used with too much faith by others can testify. We, the programmers, know that our programs are far from achieving 100% predictive power; you, the users, do not, because we are often far too good at concealing that uncomfortable fact in our papers.

Thus, having cleared my conscience, here is a chapter describing algorithms that you simply cannot survive without for doing various tricks on DNA and RNA sequences. But first we need a short background section on the molecular details of protein–DNA interactions and their role in controlling gene expression.

I. PROTEIN–DNA INTERACTIONS AND REGULATION OF GENE EXPRESSION

Many of the specific features that one can look for in a DNA sequence relate to the problem of gene regulation through protein–DNA interaction. To make matters simple, one would like to think of gene regulation in terms of signals, i.e., short stretches of defined sequence, encoded at strategic positions along the genome, that bind either regulatory proteins or polymerases. In many cases this seems to be a good conceptual model: consensus sequences can be found that are, if not always sufficient, at least necessary for the proper regulatory response, and particular regulatory proteins can sometimes be shown to bind preferentially or exclusively to these sequences. On the other hand, in many cases no consensus sequences have been found, and one has to look beyond the primary DNA sequence to understand how control is exerted. Nevertheless, the consensus sequence approach is powerful when it works, and is ideally suited for the computer biologist.

Following von Hippel and Berg (1986), site-specific protein–DNA interactions can best be dealt with by considering four different aspects of the problem: binding site specification, molecular recognition mechanisms, discrimination between the correct site and competing pseudosites, and biological expression (the regulatory response).

For a site to be specific, it must obviously be very unlikely to appear by chance in the genome in question. Thus, for *Escherichia coli,* at least 12 base pairs in the recognition site must be unambiguously specified if the expected frequency of the site reappearing at random should be less than unity [the condition $2 \times N \times (0.25)^n = 1$, where N is the size of the genome—4×10^6 bp—gives $n \approx 12$].

In reality, the recognition is not all-or-none; rather, a pattern of hydrogen bonds between the parts of the base pairs exposed in the major and/or minor grooves on the DNA and particular residues in the protein determine the specific part of the binding free energy (displacement of counterions from the phosphate backbone of the DNA may contribute a nonspecific part). Imperfect pseudosites differing in one or a few base pairs from the canonical site will compete for binding to the protein, and, although they bind with less affinities, their sheer numbers will tend to swamp the correct site in a sea of nearly correct sites. If, in addition, the number of effector protein molecules in the cell is small (the *lac* repressor is present in only 10-30 copies per cell), the probability that one of these will be bound to the target site rather than all being sequestered away on pseudosites is even further reduced.

The final regulatory response is thus seen to depend on the specification of the site (number of unambiguously defined base pairs), the mode of protein-DNA binding, the incidence of pseudosites, and the strength of the unspecific interaction, as well as on the number of effector proteins, and its degree of fluctuation from cell to cell.

The molecular structure of the protein-DNA recognition complex has been worked out in a couple of cases, notably the λ *cro* and λ repressor proteins, and the catabolite gene activator protein (CAP or CRP) (Pabo and Sauer, 1984). In these three cases, the protein binding site is built around a helix-turn-helix motif with one α-helix binding *in* the major groove and one helix lying *across* the major groove (Fig. 4.1). Residues in the helix inserted into the groove make hydrogen bonds with suitably positioned groups on the edges of the base pairs, providing the recognition specificity. Indeed, this recognition helix can be swapped between proteins with a concomitant transfer of recognition specificity (Wharton and Ptashne, 1986).

A rather different binding site has been proposed recently for deoxyribonuclease I, an endonuclease that degrades double-stranded DNA (Suck and Oefner, 1986). Here, an exposed loop fits into the minor groove, positively charged residues on either side of the loop bind electrostatically to phosphates from both DNA strands, and the DNA strand being cut interacts with two β-sheets that form the central part of the protein.

It is even possible that some DNA-binding proteins bind to DNA that is not in the standard B-form. Transcription factor IIIA, which regulates transcription of 5S RNA genes by RNA polymerase III in *Xenopus laevis,* has been suggested to bind to A-DNA, a right-handed double-helical structure with tilted base pairs similar to double-helical

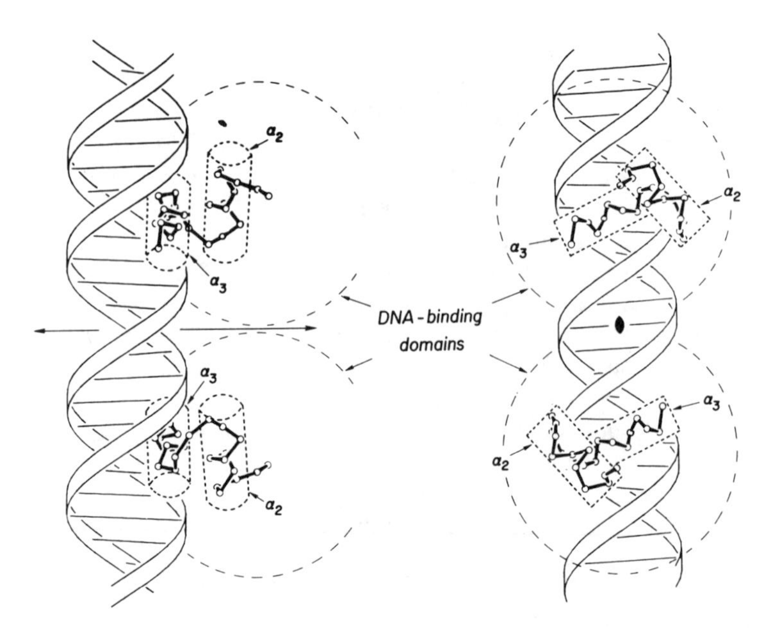

Fig. 4.1. Two views of α-helices two and three from the *cro* operator protein dimer interacting with the DNA double helix. (From Ohlendorf and Matthews, 1983. Reproduced, with permission, from the *Annual Review of Biophysics and Bioengineering,* Vol. 12, ©1983 by Annual Reviews, Inc.)

RNA (Harrison, 1986). Also, the left-handed Z-DNA structure has long been a favorite candidate for binding regulatory proteins.

Finally, not all regulatory proteins bind in close proximity to their target gene, as the so-called enhancer sequences clearly demonstrate. The mechanisms responsible for such action-at-a-distance remain largely obscure, and, in the precise vocabulary of Mark Ptashne (1986), the workers in the field neatly divide into distinct camps: the twisters (effect is propagated through changes in DNA conformation), sliders (the protein slides from one site to another), oozers (the first protein helps the next one to bind and so on until the whole stretch of DNA is covered), and loopers (proteins bound at distant sites bind to one another by looping out the intervening DNA).

II. PROMOTER SEQUENCES

Promoter sequences regulate transcription by modulating the initial recognition and binding of RNA polymerase to DNA. This can be achieved both by variations in the base sequence with which the polymerase interacts, or by (nearby or distant) sequences—operators—that bind other regulatory proteins with positive or negative influence on the polymerase-DNA interaction.

A. Prokaryotic Promoters

1. Experimental Background

Promoters from *E. coli* have been sequenced in abundance—as early as 1983 Hawley and McClure presented a list counting 112 entries—and have been a focus of interest for many workers who try to construct pattern recognition algorithms based on consensus sequences.

Biochemical studies indicate that RNA polymerase protects some 40-60 base pairs of DNA upstream and 20 base pairs downstream of the mRNA initiation site (Pribnow, 1975; Schmitz and Galas, 1979; Kammerer *et al.,* 1986), and both Pribnow (1975) and Schaller *et al.* (1975) noted a short conserved stretch of nucleotides around position −10 common to all promoters. This Pribnow box or TATA box was later shown, indeed, to be a strong consensus sequence reading TAtAaT (upper case indicates more conserved positions). The best conserved base in the −10 region is the final T, which is found in more than 95% of all promoters (Rosenberg and Court, 1979; Hawley and McClure, 1983). A second conserved region centered around position −35 was also found (Takanami *et al.,* 1976; Seeburg *et al.,* 1977); its consensus reads tcTTGACat, and in all but 12 of the 112 promoters analyzed by Hawley and McClure the spacing between the two regions was found to be 17 ± 1 bases (Fig. 4.2).

A number of mutations affecting the rate of transcription initiation have also been described. These tend to fall in the two conserved regions, and their phenotypes (increased or decreased transcription) often correlate with their fit to the consensus and to the optimal distance between the two regions (Hawley and McClure, 1983; but see also Deuschle *et al.,* 1986). It has been suggested that mutations in the −35 region primarily influence the initial binding of RNA polymerase

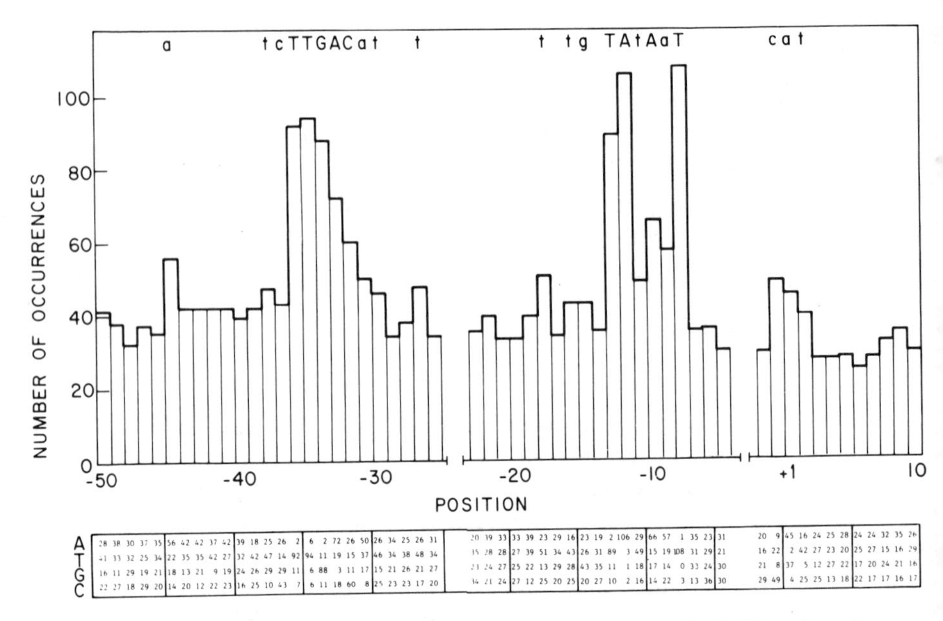

Fig. 4.2. The distribution of bases at each position in a sample of 112 prokaryotic promoter sequences. (From Hawley and McClure, 1983.) The histogram displays the number of occurrences of the most prevalent base at each position; the individual base-counts are given below. The transcription start site is base number +1. Note the conserved −10 and −35 regions.

to the promoter, whereas the −10 region seems to affect mostly the actual start of mRNA synthesis (McClure *et al.,* 1983). Regions downstream of the initiation site can also, at least in some cases, influence the promoter efficiency (Kammerer *et al.,* 1986).

Many genes that are under some kind of regulatory control have additional control elements in their promoter regions. The CAP protein, which binds cyclic AMP and serves an important role in controlling the catabolic activity of the prokaryotic cell, recognizes an upstream binding region defined by the consensus sequence TGTGA, and a second, less well-conserved, inverted repeat 6 bases downstream (de Crombrugghe *et al.,* 1984). The distance between this region and the transcription start site varies, but is typically some 50–70 bases. Interestingly, in CAP-dependent promoters the −35 and −10 regions show only poor fits to the canonical consensus, and the spacing is often far from optimal. Any constitutive expression would thus be expected to be at a very low level, and the effector molecule can exert its full regulatory potential. Other coordinately regulated *E. coli* promoters also have conserved features not normally present in unregulated ones, such as a GC-rich sequence abutting onto the transcription start point in promoters under stringent control (Travers, 1984), or an altered −10 region in heat shock genes (Cowing *et al.*, 1985).

2. *Prediction Schemes and Other Theoretical Analyses*

The Hawley and McClure (1983) compilation of promoter sequences has been used repeatedly as a basis for prediction algorithms. Mulligan *et al.* (1984) designed a typical program for picking out likely promoter sites working as follows: First, the locations of sequences homologous to the −35 and −10 consensus sequences (TTGACA and TATAAT) are found by requiring at least three matches out of six for both signals. Second, only those matches where the spacing between a putative −35 and a putative −10 region is between 15 to 21 base pairs are retained. Third, these candidates are evaluated according to a weighting scheme. To this end, a weight matrix is constructed by noting the incidence of each residue in each position in the alignment of Fig. 4.2, and normalizing by the square root of the expected incidence in a random sample ($\sqrt{112/4} = 5.3$ when the frequencies of the four bases are assumed equal). The candidate sequence is then matched against the weight matrix by summing its particular weights as they are read off from the matrix (including a contribution from the observed spacing with weights 14 for a 17-base spacing, 6 for 16- and 18-base spacings, and 1 for 15- and 19- to 21-base spacings), and a homology score is calculated from the equation:

$$\text{Score} = 100[(\text{sum of base pair weights} + \text{spacing weight} - \text{baseline weight})/(\text{maximum weight for perfect match} - \text{baseline weight})] \quad (4.1)$$

where the baseline weight is the expectation for a random occurrence of all four bases. Finally, if the score is higher than a cutoff value of 45, the candidate is considered a likely promoter.

In the weight matrix, a few weak signals in the known promoters beyond the −35 and −10 hexamers are included by extending it 9 bases upstream and a single base downstream of the −35 consensus, and 5 bases upstream and 3 bases downstream of the −10 region.

On the basis of an extensive test of the method on both random sequences and real sequences from GenBank, it was estimated that the number of false negatives (real promoters missed by the program) is small (0–10%), whereas the number of false positives (good candidates that in reality do not have promoter activity) is rather high, maybe as much as 50% (Mulligan and McClure, 1986). The authors also point out that the overall base composition of the genome in question has a profound influence on the number of candidates that the program will find (more candidates as the AT content goes up), and suggest that a higher cutoff value should be used when the AT content is high. Interestingly, there is a fairly good correlation between the homology score and promoter strength, such that the rate of transcription initiation can be predicted to within a factor of 4 over a range of 10^4 (Mulligan *et al.*, 1984).

Recently, Stormo *et al.* (1986) have extended this method by showing how a weight matrix can be optimized with respect to its ability to predict the activity of a given signal; this is achieved by treating the matrix elements as unknown variables, calculating the score in terms of these variables for all signals in the data set, and solving the resulting (overdetermined) set of linear equations (score = activity) by multiple linear regression.

Staden (1984a) has developed another algorithm similar to Hawley and McClure's (1984). The main differences are that Staden bypasses the first search for three-out-of-six matching, and calculates the score by *multiplying* rather than summing (as in the previous method) the frequencies with which the bases in the candidate sequence appear in the weight matrix, thus arriving at a (theoretically more pleasing) measure of the probability that the section scanned is in fact a promoter. Staden also includes a weight for the spacing, and adds a score for the region around the mRNA start site (positions −2 to +10) where some base preferences exist (Fig. 4.2). In practice, the two methods make very similar predictions, and are clearly helpful when making a preliminary characterization of a new DNA sequence. However, in the words of Staden (1984a), "Gene search by content-methods (i.e. methods based on codon bias, section VI.B) . . . are, so far, more likely to correctly indicate coding regions than any of the gene search by signal-methods."

Rather than trying to predict new promoters, some workers have studied the Hawley and McClure alignments from the point of view of information content and protein–DNA recognition. Using information theory, Schneider *et al.* (1986), in an attempt to measure the degree of nonrandomness of a sequence, have devised a method for calculating the number of bits needed to define a DNA signal such as a promoter. For a given position n in the signal they define the uncertainty as:

$$H_s(n) = -\Sigma f_{i,n} \log_2 f_{i,n} \text{ (bits per base)} \qquad (4.2)$$

where $f_{i,n}$ is the frequency of base i and the summation is over the four bases (i = A,C,G,T).

For a random sequence with all four bases equally represented (as in *E. coli*), the uncertainty is $H_g = 2$ bits per base. The nonrandomness in position n can thus be measured by $R(n) = H_g - H_s(n)$, and the total information content of the whole signal is $R_{tot} = \Sigma R(n)$ with summation over all sites in the signal. R_{tot} as calculated from alignments such as in Fig. 4.2 can then be compared with the minimum number of bits it takes to specify a unique site in the whole genome of *E. coli* (with an estimated size $N = 4 \times 10^6$ bp), $R_{genome} = \log_2(2N) = 22.9$ bits. Typical repressor binding-sites turn out to have $R_{tot}/R_{genome} \approx 1$, but for the promoters R_{tot} is only 11.1 bits, corresponding to a very large number of potential sites in the whole genome (one per 1000 bp). It is thus not surprising that the prediction methods based on the Hawley and McClure compilation yield a large number of false positives.

A related approach has recently been taken by Berg and von Hippel (1987) with the ambitious aim of trying to find a relation between sequence variability within a signal such as the promoter and the protein–DNA interaction free energy contributed by each base pair in the signal. Noting a formal analogy between the microcanonical ensemble in statistical physics and the allowed sequence variability in a site with a discrimination energy in some range around a required level, they derive equations that make it possible to predict the binding free energy and the strength of any promoter sequence, given a large enough data base of known promoters. In practice, and given the current size of the data base, this method is about as accurate as that of Mulligan *et al.* (1984) for predicting promoter strengths.

As an added bonus, the information theory-based analysis of Schneider *et al.* appears naturally in the Berg–von Hippel theory as the entropic aspect of signal specificity, but only in the latter can one estimate the contributions to the actual binding free energy from individual base pairs. This last method is thus more rigorous and potentially (given enough sequences) more powerful in extracting the maximum amount of information from the data, but is also more complicated than the straightforward matrix or information theory methods.

Throughout this section, it has been taken for granted that the initial aligning of the sequences is uncomplicated, and this is indeed the case when one is dealing with signals as distinct as the *E. coli* promoters. It would nevertheless be nice to have a general alignment algorithm that could be used to detect weaker signals without any prior knowledge of what to expect. A method that promises to do just this, given only a very approximate initial alignment, has been devised by Galas *et al.* (1985). In essence, this algorithm searches for the occurrence of all k-letter words in a window of size W that is moved along the set of approximately aligned sequences (a large W means that signals even in poorly aligned sequences will be picked up). To each k-letter word is associated a set of neighbors, i.e., words that differ from it by no more than d letters. For any word w found in any given window position, a score is calculated by adding all occurrences of w and its neighbors weighted according to the number of letters in common with w. The score for the most common word w' is then plotted for each position of the window (Fig. 4.3) and any strong consensus patterns are immedi-

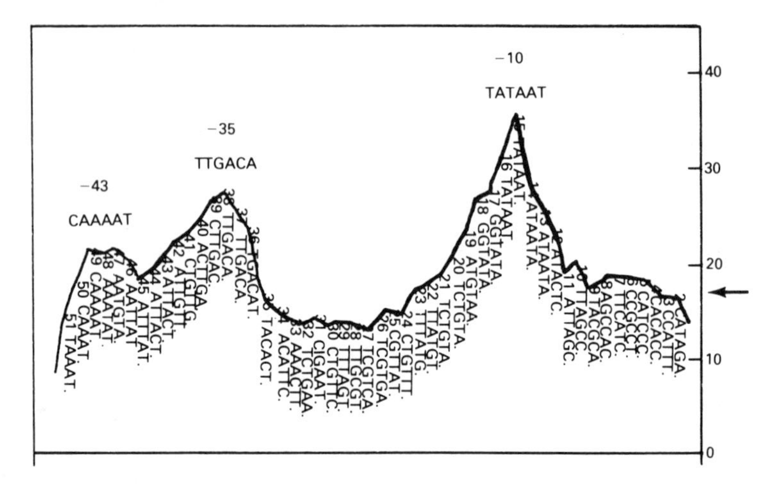

Fig. 4.3. Score for the highest-scoring k-letter word ($k = 6$) as a function of window position (window-length = 9 and maximum number of mismatches $d = 2$) for a sample of prokaryotic promoters. The highest-scoring word at each position is also shown, and the expected value for random sequences is marked (arrow). The sequences are aligned on the transcription start site. Cf. Fig. 4.2. (From Galas *et al.*, 1985.)

ately seen. One can then realign the sequences on any one of the patterns thus identified, and finally arrive at an optimal alignment.

In the case of the promoter sequences, and starting with an alignment from the transcription start site, this method easily identifies the −35 and −10 regions, together with a weaker region around position −44 that may be of functional significance. The strength of this approach clearly lies in its ability to find all or most conserved regions (signals) in a set of approximately aligned sequences in only a small number of runs through the data.

B. Eukaryotic Promoters

1. *Experimental Background*

Eukaryotic promoters are much more complex than their prokaryotic counterparts in that the number of different regulatory signals affecting transcription is much larger. Nevertheless, the idea of short, specific sequence motifs serving as binding sites for regulatory proteins seems to hold quite well, and gene regulation is usually portrayed as involving plugging one or more casettes into the promoter region.

A particularly clear example of this is shown in Fig. 4.4, where the results from a fine-structure mutational analysis of a β-globin promoter transcribed by RNA polymerase II are presented (Myers *et al.*, 1986). Three previously identified regions seem to confer most of the specificity in this system: a β-globin-specific region at about position −90, the commonly found CCAAT box at −75, and the ubiquitous TATA or Hogness–Goldberg box some 30 residues upstream from the site of transcription initiation.

Other genes have been shown to have other specific boxes in their upstream regions. Thus, a regulatory factor called Sp1 binds to the element GGGCGG (Dynan and Tjian, 1985; Briggs *et al.*, 1986); so-called heat shock genes are regulated by a heat shock transcription factor that recognizes a self-complementary consensus sequence C--GAA--TTC-G (Pelham, 1985); and the transcription of many genes is stimulated by the presence of enhancers that can exert their effect over distances measured in thousands of bases and irrespective of their orientation (Serfling *et al.*, 1985; Zenke *et al.*, 1986). In fact, RNA polymerase II cannot by itself initiate transcription; it needs additional factors, at least *in vitro* (Dynan and Tjian, 1985).

Even though the distances between the casettes can vary, the stereo-

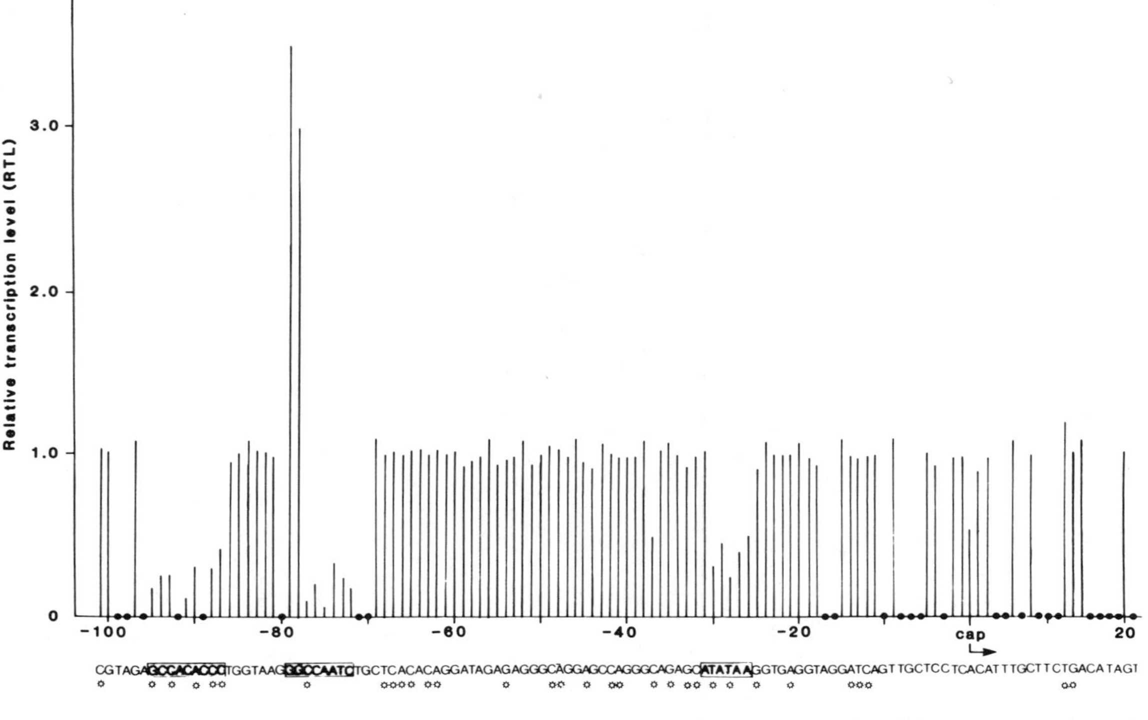

Fig. 4.4. Relative transcription levels (wild type = 1) of mutant β-globin promoters. The wild-type sequence is on the abscissa. Solid circles, positions where no mutants were obtained; open asterisks, positions where multiple mutations were isolated. Conserved promoter elements are boxed. (From Myers *et al.*, 1986. Copyright 1986 by the AAAS.)

specific alignment of them on the surface of the DNA helix appears to be crucial. Half-turn but not full-turn deletions and insertions between the enhancer and another gene-specific signal in the simian virus 40 early promoter, as well as between the specific signal and the TATA box, seriously cut down on transcription from this promoter (Takahashi *et al.,* 1986). Changing the angular separation between the TATA box and the mRNA initiation site (the Cap site) in a protamine gene likewise affects the precise position of initiation (Kovacs and Butterworth, 1986).

Other eukaryotic RNA polymerases such as RNA polymerase III that transcribe 5S RNA, tRNA, and some small RNAs also need regulatory protein cofactors to initiate transcription, and are activated by conserved casettes in their target DNA sequences (Ciliberto *et al.,* 1983)—remarkably, the regulatory region is found in the coding sequence downstream from the initiation site in this case. One of the factors conferring specificity onto RNA polymerase III, the so-called transcription factor IIIA from *Xenopus,* apparently has a modular design with nine repeated domains or fingers (Miller *et al.,* 1985), and binds to short runs of guanines at about 5-nucleotide intervals in the control region (Rhodes and Klug, 1986).

2. *Theoretical Analyses*

The panoply of different regulatory sequences in the eukaryotic promoters has made theoretical analysis difficult, since only a few examples of any given signal are known. The TATA box is well established, however, and, as shown in Fig. 4.5, is embedded in a region of relatively low AT content (Nussinov *et al.,* 1986; Bucher and Trifonov, 1986). The CCAAT box is much less distinct (not shown), but a second TATA-like region appears around position −275 in both mammalian and invertebrate sequences.

So far, no reliable schemes for finding eukaryotic promoters have

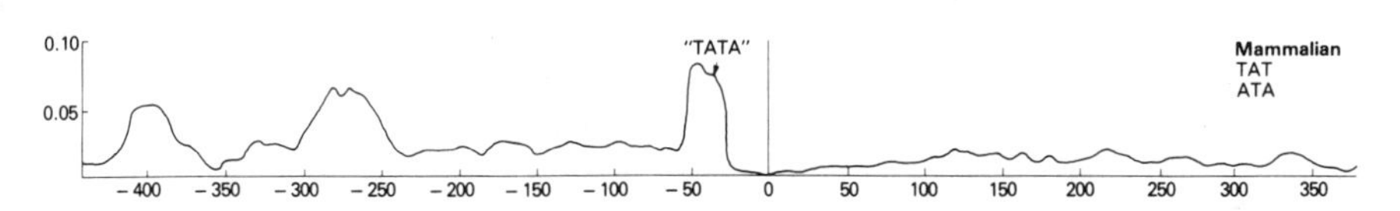

Fig. 4.5. The distribution of ATA + TAT triplets around transcription initiation sites ($x = 0$) in 139 mammalian genes (Nussinov *et al.,* 1986). The y value is the ratio of the triplet count in a 25-residue-long window to the total number of bases in this window ($= 25 \times 139$).

been forthcoming. Indeed, Sadler *et al.* (1983) have shown that the CCAAT and TATA boxes are far from sufficient to locate promoter sites, again stressing the point that the RNA polymerases seem to need additional, gene-specific factors to initiate transcription.

III. TERMINATOR SEQUENCES

If initiation of transcription can be constitutive as well as regulatable (factor dependent), the same is true for termination. Constitutive termination is so far the better understood of the two, and prokaryotic factor-independent terminators can be detected by computerized algorithms with a fair degree of reliability. Some strong signals associated not with the actual termination event but rather with the endonucleolytic cleavage and poly(A) addition that generate the mature RNA terminus in eukaryotic mRNAs have also been defined, but they are as yet not sufficiently well understood to allow unambiguous predictions to be made. In a recent review, Platt (1986) summed up our present level of ignorance by stating that the answers to the two questions, "What nucleic acid sequence(s) or structure, if any, constitute a signal for transcription termination?" and "How does the transcriptional apparatus recognize and utilize the termination signal?" are "still dismayingly vague." Vague as they may be, this has (un?) fortunately not completely deterred the signal-hunters and prediction-spawners from doing their thing.

A. Prokaryotic Terminators

1. Experimental Background

The classical constitutive prokaryotic terminator is composed of a GC-rich dyad symmetry centered some 12–24 bases upstream of the termination site, immediately followed by a run of about 8Ts (Rosenberg and Court, 1979; Platt, 1986; Brendel *et al.,* 1986). Studies of mutations, deletions, and the effects of incorporating base analogs in the RNA have suggested that the dyad symmetry is realized in the form of a rather stable RNA hairpin loop that induces a transcriptional pause during which the exceptionally labile rU-dA RNA-DNA hybrid dissociates, thereby terminating transcription (Farnham and Platt, 1980).

The typical nonconstitutive terminator is the *rho*-dependent one, which apparently does not involve any easily recognizable DNA signal. Rather, a long stretch of upstream sequence (some 80 bp) seems to be required for efficient termination by *rho,* but it is not obvious what the important features of this region are (Platt, 1986). Other, less well-studied, termination factors include the *nusA* protein and the *tau* factor (Platt, 1986; Briat and Chamberlin, 1984).

2. *Prediction Schemes*

Brendel and collaborators have published two methods for detecting constitutive terminators in prokaryotic DNA (Brendel and Trifonov, 1984; Brendel *et al.,* 1986). The weight matrix approach described in the section on promoters above is used here as well, but now *di-* rather than mononucleotide frequencies are tabulated. The stated reason for this choice is that structural and physical properties beyond the base-per-base pattern might be of importance in protein–DNA interactions; one such set of parameters could be the local twist-and-roll angles in the DNA helix (Section VIII,A) that depend primarily on base-stacking interactions, hence on dinucleotide patterns. It would be interesting to compare the results from mono- and dinucleotide weight matrices, but no such comparison has been published.

To construct the weight matrix, 51 known terminators were initially aligned from their termination sites (i.e., the 3′ ends of the RNAs). However, since the point of termination is not always precisely defined, each sequence was realigned to the matrix derived from the other 50 sequences, and the process was repeated until no more shifts were observed. The weight matrix was thus in a sense internally optimized in a way reminiscent of the method used by Galas *et al.* (1985) to find consensus sequences described in Section II,A,2 above. Incidentally, two hitherto unrecognized consensus patterns became apparent in this compilation, namely CGGG(C/G) upstream and TCTG downstream of the run of Ts.

The number of dinucleotides being four times the number of bases, the weight matrix (which includes positions −42 to +8) looks rather formidable (Fig. 4.6). To reduce the noise level, all frequencies not significantly different from the random expectation were put equal to zero. The match p of a given 50-bp sequence to the matrix is obtained by summing the respective matrix elements as one dinucleotide after another is read off from the sequence, and the resulting score is normalized by subtracting the average score for random sequences of *E. coli*

			-40					-35					-30					-25					-20					-15					-10					-5				-1	+1				+5				
AA	o	o	o	1	1	o	o	1	1	1	1	o	o	1	1	o	o	o	o	o	o	o	o	1	o	o	o	o	o	o	o	o	o	o	o	o	o	o	o	o	o	o	o	o	o	o	o	o	1	1	AA
AC	o	o	o	o	o	o	o	o	o	o	.43	.46	o	o	o	o	o	o	o	o	o	o	o	o	o	o	o	o	o	o	o	o	o	o	o	o	o	o	o	o	o	o	o	o	o	o	o	o	o	o	AC
AG	1	o	o	o	o	o	o	.58	o	o	o	1	o	o	o	.75	o	o	o	o	o	o	o	o	o	o	o	o	o	o	o	30	o	o	o	o	o	o	o	o	o	o	o	o	o	o	o	o	o	o	AG
AT	o	1	1	o	o	o	o	o	o	.73	o	o	o	o	1	o	o	o	o	o	o	o	o	o	1	o	o	o	o	o	o	o	o	o	o	o	o	o	o	o	o	.32	o	.67	o	o	o	o	o	o	AT
CA	o	.78	o	o	o	o	1	o	o	.73	.50	o	o	o	o	o	o	o	o	1	o	o	o	o	o	o	o	o	o	o	o	o	o	o	o	o	o	o	o	o	o	o	o	o	o	o	o	o	o	o	CA
CC	o	o	o	o	o	o	o	o	o	o	o	o	1	.71	.88	o	o	1	1	.78	o	1	o	o	o	o	o	o	.88	o	o	o	o	o	o	o	o	o	o	o	o	o	o	o	o	o	o	o	o	o	CC
CG	o	o	o	o	o	o	o	.58	o	o	o	o	o	o	o	o	o	o	o	1	o	o	o	o	o	o	.58	o	o	1	.56	.30	o	o	o	o	o	o	o	o	o	o	o	o	o	o	o	o	o	o	CG
CT	o	o	o	o	o	o	o	o	o	o	o	o	o	o	o	o	o	.70	o	o	o	o	1	o	o	o	o	.78	o	o	o	o	o	.73	.75	o	o	o	o	o	o	o	o	o	o	1	o	o	o	o	CT
GA	o	o	o	o	o	1	o	o	.88	o	o	o	.75	o	o	o	o	o	o	o	o	o	o	o	o	o	o	o	o	o	o	o	o	o	o	o	o	o	o	o	o	o	o	o	o	o	o	.60	o	o	GA
GC	o	o	o	o	o	o	o	o	o	o	o	o	o	.50	o	o	1	o	o	o	1	o	o	o	o	o	o	o	1	.53	.39	o	.61	.82	o	o	o	o	o	o	o	o	o	o	o	o	o	1	o	o	GC
GG	o	o	o	o	o	o	o	o	o	o	o	o	o	o	o	1	o	o	o	o	o	o	o	o	o	o	o	o	o	.60	1	1	1	1	o	o	o	o	o	o	o	o	o	o	o	o	o	o	o	o	GG
GT	o	o	o	o	o	o	o	o	o	o	o	o	o	o	o	o	o	o	o	o	o	o	o	o	o	o	o	o	o	o	o	o	o	o	1	o	o	o	o	o	o	o	o	o	o	o	o	o	o	o	GT
TA	o	o	o	o	.64	o	o	o	o	o	.43	o	o	o	o	o	o	o	o	o	o	o	o	o	o	o	o	o	o	o	o	o	o	o	o	o	o	o	o	o	.25	o	1	o	o	o	o	o	.88	o	TA
TC	o	o	o	o	o	o	o	o	o	o	o	o	o	o	o	o	o	o	.55	o	o	o	o	o	.55	o	o	1	o	o	o	o	o	o	o	o	o	o	o	o	o	o	o	o	1	o	o	o	o	o	TC
TG	o	o	o	o	o	o	o	o	o	o	o	o	o	o	o	o	o	o	o	o	o	o	o	o	o	o	o	o	o	o	o	o	o	o	o	o	o	o	.16	.17	o	o	.71	o	o	o	1	o	o	o	TG
TT	o	o	o	o	o	o	o	o	o	o	o	o	o	o	o	o	o	o	o	o	o	o	o	o	o	o	1	o	o	o	o	o	o	o	.92	1	1	1	1	1	1	1	.79	1	o	o	o	o	o	o	TT

C C A A T A T T A C A T A A C A A T C C T C G C A C T C G C G G G G A T T T A T T T T A T C T G A A C
0 .78 0 0 .64 0 0 0 0 .73 0 0 0 0 0 0 0 0 1 0 0 0 0 0 0 0 0 0 1 1 1 1 1 0 0 1 1 0 0 1 1 1 1 .67 1 1 1 .60 1 0

p' = 0 + .78 + ... = 19.42 ;   p = (19.42 − 4.8) / 2.23 = 6.56

Fig. 4.6. Normalized dinucleotide distribution matrix for 51 prokaryotic terminators (the termination site is between positions −1 and +1). A calculation of the *p* value for a putative termination site is shown below (see text). (From Brendel *et al.*, 1986.)

base composition and dividing by the standard deviation for random sequences.

This weight matrix alone proved to perform reasonably well, missing some 5–6% of the known terminators while passing only 0.5% of random sequences with $p > 3.0$ as a cutoff (Brendel and Trifonov, 1984). An improvement was obtained recently by introducing a second weight matrix representing not sequence per se but the degree of dyad summetry in the hairpin region (Brendel *et al.*, 1986). In setting up this matrix, only the standard dyad symmetry was included; thus, only potential hairpins meeting the following criteria were considered: the center of the hairpin lies between −24 and −12; the loop size is from 3 to 7 bp; the stem is closed at both ends by at least two base pairs; the terminal base pairs are not G-U; the stem length is at least 4 bp; G-U base pairs comprise less then 15% of the total; and at least 75% of the stem base pairs are G-C, A-T, or G-U. With these rules, a complementarity weight matrix (Fig. 4.7), can be constructed from the 51 known terminator sequences; the standard dyad symmetry is apparent as a diagonal of relatively large numbers. The match s between the matrix and any test sequence is calculated by first constructing the comple-

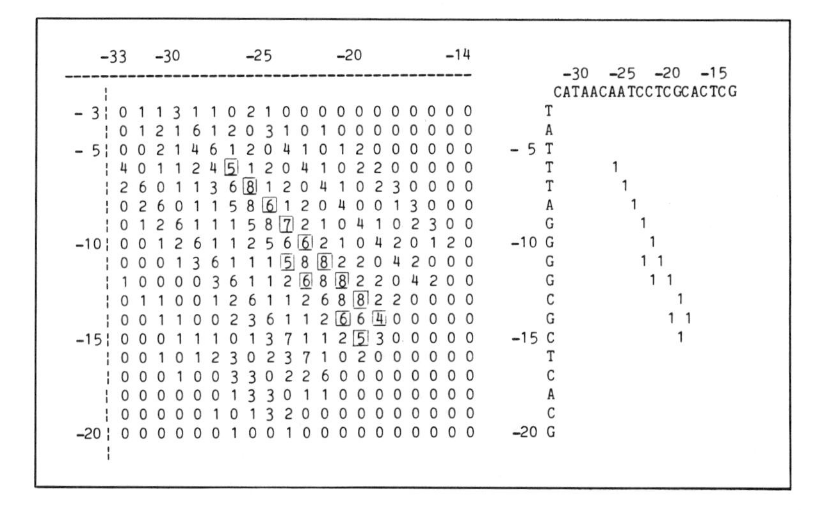

Fig. 4.7. Complementarity weight matrix for 51 prokaryotic terminators, i.e., the sum of complementarity diagrams such as shown on the right for the whole sample. The s value for the right-hand matrix is obtained by adding (i) the number of matches in the matrix being analyzed (= 13) and (ii) all values from the weight matrix corresponding to these matches (boxed); in this particular case, $s = 95$. (From Brendel *et al.*, 1986.)

mentarity diagram for the test sequence using the above rules (except the first), summing the number of complementarities found in this diagram, and finally adding all elements from the full complementarity matrix that correspond to matches in the diagram for the test sequence. Thus, even though the complementarity pattern in the test sequence may not correspond exactly to the standard one, a strong dyad symmetry will nevertheless count somewhat.

The optimal thresholds for predicting terminators, now using both the p and s scores, are either $p > 3.5$ or both $p > 3.0$ and $s > 5.0$. With these cutoffs, 49/51 terminators were identified correctly, while less than 0.5% of random sequences qualify. The improvement over the earlier method is but slight, however: the s value varies a lot more than the p value between different terminators, making it a less efficient predictor.

B. Eukaryotic Terminators

1. Experimental Background

As for the eukaryotic promoters, no algorithm for finding eukaryotic terminators has yet been published, and the sequence determinants are not at all well understood. A major complication is that termination proper usually takes place a considerable distance downstream from the final mature 3′ terminus of the RNA; hence any patterns found in the RNA are likely to be involved in posttranscriptional processing events (endonucleolytic cleavage and poly(A) addition) rather than in termination per se.

Nevertheless, RNA polymerase III (which transcribes 5S RNA, tRNA, and some small RNAs) most often terminates at runs of four or more Ts (somewhat longer runs in yeast), although it can also recognize other signals such as long runs of A residues (Platt, 1986). The sequences specifying termination for RNA polymerase I (transcribes ribosomal RNAs) are even less well characterized.

Since most mRNAs have polyadenylated 3′ ends, it has been easier to study the posttranscriptional processing subsequent to RNA polymerase II termination rather than the termination event itself. No obvious termination signals are known except in yeast, where the 8-bp consensus TTTTTATA has been implicated as part of the necessary determinant—but, alas, other downstream elements still seem to be required for full effectiveness (Henikoff and Cohen, 1984; McLauchlan *et al.*, 1985; McDevitt *et al.*, 1986). The poly(A) signal AAUAAA (Proudfoot and Brownlee, 1976), which is normally found some 11–30 residues upstream of the poly(A) tail (Fitzgerald and Shenk, 1981), is a hallmark of eukaryotic mRNA 3′ termini, and is indeed well conserved. Some variation in the first two positions (UAUAAA and AUUAAA) is possible, whereas the remaining four bases seem more critical; AAGAAA, AAUAAG, and AAUACA do not function properly (Mason *et al.*, 1985). When more than one possible signal is present, the most upstream one is the one used most frequently, either because the cleavage/polyadenylation machinery scans the RNA in the 5′→3′ direction, or because simultaneous processing at all possible sites will result in a molecule with only the most upstream one left (Mason *et al.*, 1985). The AAUAAA signal cannot be solely responsible for 3′ processing, however, since it will appear by chance about once every 2000 bp, and indeed is found in the coding region of some genes. Again, downstream elements appear to be important for specifying the correct site (Platt, 1986), and, conversely, an intact AAUAAA signal may be required for effective termination (Whitelaw and Proudfoot, 1986). Current thinking on the events involved in termination and processing are summarized in Fig. 4.8.

2. *Theoretical Studies*

The paucity of clearly identified DNA signals in this area is mirrored by a paucity of theoretical studies. The poly(A) signal is easily identified, and was in fact suggested on the basis of only six sequences (Proudfoot and Brownlee, 1976), but beyond this not much progress has been made. Berget (1984) has proposed a CAPyUG recognition element close to the point of poly(A) addition, that, together with the AAUAAA sequence, might bind to complementary regions in the small nuclear (sn) RNA U4 that is believed to play a role in the process.

The most exhaustive study to date is Nussinov's (1986) analysis of 216 sequences from GenBank, where she noted the occurrence of all 256 base quartets as a function of the distance from the site of poly(A) addition. Her most significant results concern the low incidence of the component quartets from the AAUAAA signal upstream of the poly(A) site, a clustering of T residues just downstream of this site, and an increase in the number of alternating GT oligomers immediately surrounding the poly(A) site (Fig. 4.9). One problem with this kind of

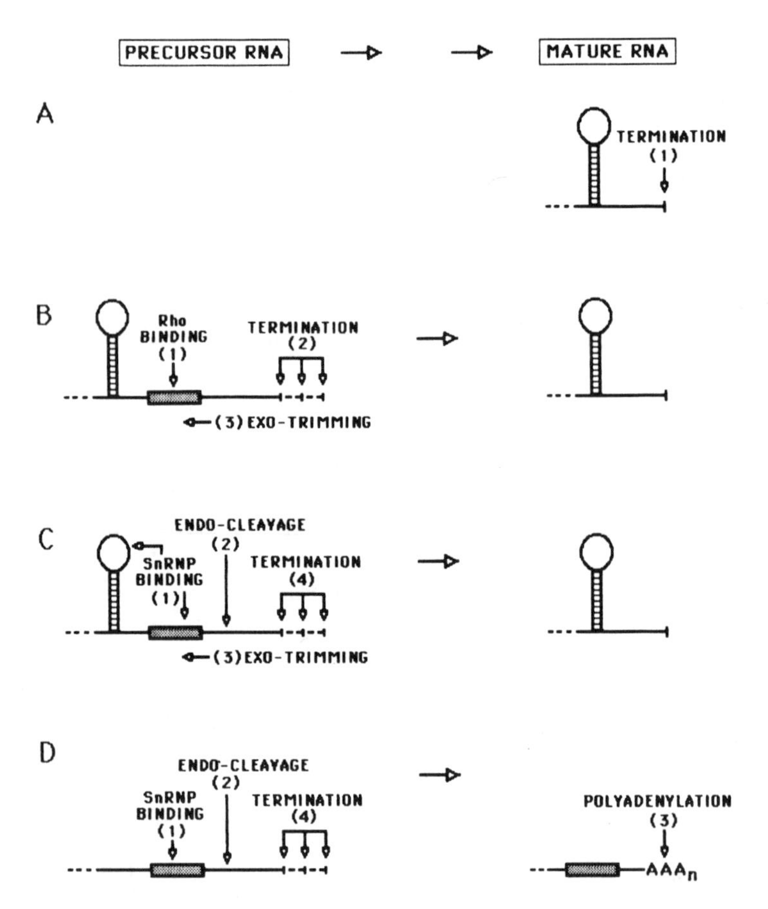

Fig. 4.8. Mechanisms for the generation of RNA 3′ ends. In prokaryotes, this may involve simple termination (A), or termination followed by nuclease trimming (B). In eukaryotes, hairpins may (histone mRNAs, C) or may not (D) be involved in defining the 3′ end. (From Platt, 1986. Reproduced, with permission, from the *Annual Review of Biochemistry,* Volume 55, © 1986 by Annual Reviews, Inc.)

analysis is that the statistical significance of the results is very hard to estimate.

From the point of view of actually finding termination sites and predicting the most likely position of the mRNA 3′ terminus, we are thus left with not much else to entertain us with but to look for AAUAAA signals downstream of the presumed termination codon, and hope that the most 5′ of these is in fact the one used *in vivo.* A search routine of this kind is included in, e.g., Staden's package ANALYSEQ (Staden, 1986).

IV. SPLICE SITES

Ever since the discovery of split genes, now almost a decade ago, the nature of the splicing reaction as well as the evolutionary background of this extraordinary phenomenon has been the center of an intense research effort. Nevertheless, as we shall see, totally reliable predictions of splice junctions based only on gene sequences (not invoking cDNA data) are possibly within reach but not yet at hand. Theoretical sequence analyses have of course provided consensus sequences of var-

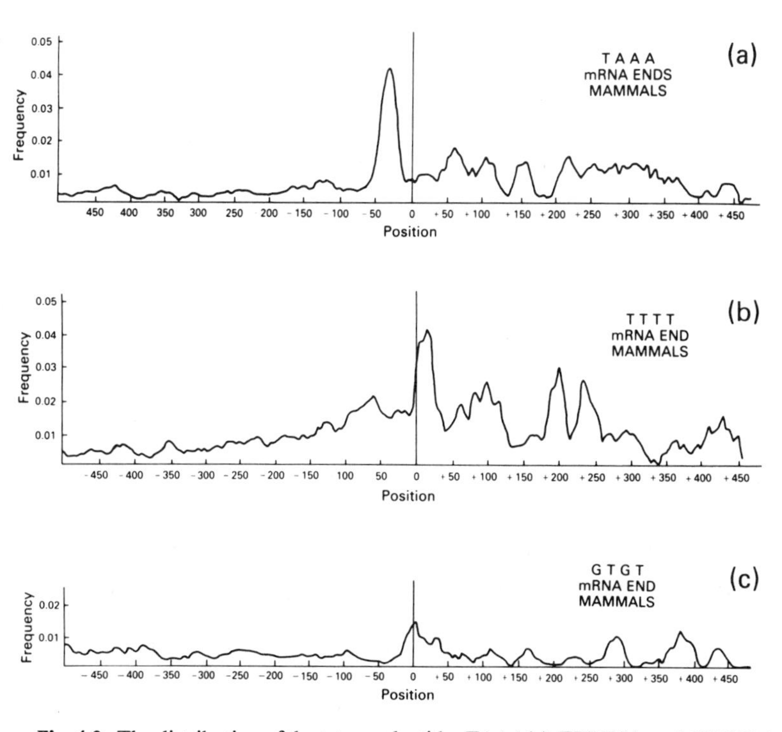

Fig. 4.9. The distribution of the tetranucleotides TAAA(a), TTTT(b), and GTGT(c) around the 3′ ends of mammalian mRNAs (216 sequences). The frequency on the y axis is the number of occurrences of the tetranucleotide in a 25-residue window divided by the number of nucleotides in the window ($=25 \times 216$). (From Nussinov, 1986.)

ious kinds—the lariat branch point sequence was in fact detected as a conserved element close to 3′ splice sites before the lariat intermediate had been discovered experimentally—and, equally important, have served to illuminate the evolutionary aspects of splicing. It has thus been shown that splice junctions tend to map at protein surfaces (Craik *et al.*, 1982), between compact folding units such as $\beta\alpha\beta$ patterns (Gō, 1981, 1983a), or between transmembrane elements in many integral membrane proteins (Argos and Rao, 1985). The sizes of exons (the regions of the gene retained in the mRNA) tend to be relatively sharply distributed around 40–50 amino acid residues (120–150 bases), whereas introns (the regions spliced out) come in all lengths ranging up to 10,000 bases or more.

The hypothesis that exons should correspond to structural and/or functional units of proteins has thus received much support, and mosaic genes composed of an assortment of exons from a variety of other genes have recently been found (Gilbert, 1985). The prevailing view at present is that split genes were present in the common ancestor of prokaryotes and eukaryotes; that the streamlined prokaryotes have done away with the introns to save on DNA and protein; and that eukaryote gene evolution proceeds through exon shuffling as well as through intron loss, intron sliding, and, possibly, intron insertion (Gilbert, 1985; Rogers, 1985; Gilbert *et al.*, 1986).

A. Experimental Background

Up until 1984, focus was on the 5′ and 3′ splice sites and their immediate surroundings: the so-called GT-AG rule for splice junctions was formulated (Breathnach *et al.*, 1978; Breathnach and Chambon, 1981), and extended to take further consensus patterns into account (Mount, 1982). With the advent of efficient *in vitro* splicing systems, a new twist was added to the story when the lariat intermediate was discovered (Keller, 1984). Thus, splicing is now believed to proceed through a lariat form with the 5′ G of the intron covalently attached to the 2′ position of an A residue contained within a weakly conserved consensus sequence 18–40 bases upstream of the 3′ intron–exon junction (Fig. 4.10). This intermediate is then further processed to yield a free lariat intron and an intact RNA with the correct exon–exon splice junction.

Point mutations in the 3′ splice site block splicing but not 5′ cleavage and lariat formation; removal of a longer segment including the 3′ consensus sequence but not the lariat branch point blocks 5′ cleavage

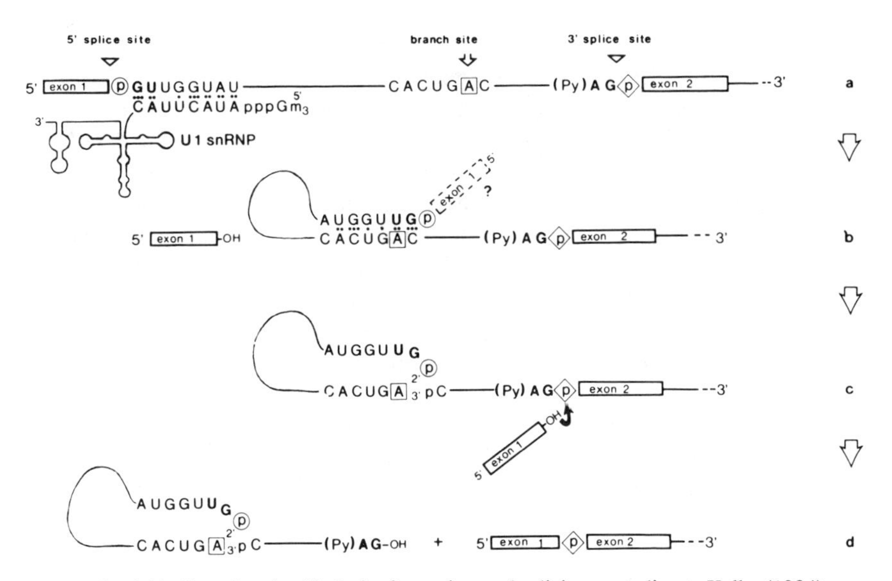

Fig. 4.10. Steps involved in lariat formation and splicing according to Keller (1984).

as well; and cryptic lariat acceptors can be activated by removing the normal site (Reed and Maniatis, 1985; Aebi *et al.,* 1986). It has also been shown by introducing competing splice sites in close proximity that the more like the consensus sequence (Section IV, B) a site is, the more efficient it is as a substrate (Eperon *et al.,* 1986).

Snurps, i.e., small nuclear ribonucleoprotein particles (snRNPs), have been directly implicated in the 5′ cleavage reaction (Furdon and Kole, 1986; Zhuang and Weiner, 1986), to some extent substantiating earlier suggestions based on sequence complementarity between the U1 snRNA and the 5′ and 3′ splice junctions (Lerner *et al.,* 1980; Avvedimento *et al.,* 1980; Oshima *et al.,* 1981).

Finally, it has been proposed that splicing, at least in yeast, takes place on a 40 S spliceosome (how many "some's" do we have in molecular biology?) that holds the free end of the 5′ exon generated in the first step of lariat formation in a position such that it can be ligated to the 3′ exon in the subsequent step (Brody and Abelson, 1985).

B. Theoretical Analyses and Prediction Methods

The yeast consensus sequences are highly conserved: A hexanucleotide GTAPyGT at the 5′ end of the intron and the so-called TACTAAC box some 6–60 bases upstream of the 3′ AG of the intron are invariably present (Langford and Gallwitz, 1983), although neighboring sequences may also have some effect on the efficiency of the splicing reaction (Cellini *et al.*, 1986). Lariat formation involves the 5′ G of the intron and the A in position six of the TACTAAC box (Domdey *et al.*, 1984).

In the higher eukaryotes, similar conserved regions can be found, but they are much more variable. An extensive compilation of splice junction sequences (Mount, 1982) (Fig. 4.11) reveals a 5′ consensus (C,A)AG/GT(A,G)AGT, and a 3′ consensus $(T,C)_n N(C,T)AG/G$ where $n \geq 11$ (the slash indicates the position of the intron–exon boundary). A more refined search for consensus sequences upstream of 3′ splice junctions reminiscent of the method of Galas *et al.* (1985) described in Section II,A,2, but searching only for sequence more or less homologous to the yeast TACTAAC box found a C(G,A)A(T,C) pattern around the lariat branch point in RNA from higher eukaryotes (Keller and Noon, 1984). This sequence, together with the 3′ consensus, was proposed to make complementary base pairs with the U2 snRNA.

Donor Sequences

position	-4	-3	-2	-1	+1	+2	+3	+4	+5	+6	+7	+8
%A	30	40	64	9	0	0	62	68	9	17	39	24
%T	20	7	13	12	0	100	6	12	5	63	22	26
%C	30	43	12	6	0	0	2	9	2	12	21	29
%G	19	9	12	73	100	0	29	12	84	9	18	20

Acceptor Sequences

	-15	-14	-13	-12	-11	-10	-9	-8	-7	-6	-5	-4	-3	-2	-1	+1	+2	+3
%A	15	10	10	15	6	15	11	19	12	3	10	25	4	100	0	22	17	19
%T	51	44	50	53	60	49	49	45	45	57	58	29	31	0	0	8	37	28
%C	19	25	31	21	24	30	33	28	36	36	28	22	65	0	0	18	22	32
%G	15	21	10	10	10	6	7	9	7	4	5	24	1	0	100	52	25	20

Donor Consensus: (C,A)AG/GT(A,G)AGT

Acceptor Consensus: $(T,C)_{11}N(C,T)AG/G$

Fig. 4.11. Distribution of nucleotides around the 5′ donor (139 sequences) and 3′ acceptor (130 sequences) splice sites. (From Mount, 1982.)

Turning now to actual prediction schemes, Staden (1984a) has devised a weight matrix method analogous to his promoter search algorithm described in Section II,A,2. Using the weights in Mount's compilation (Fig. 4.11), separate scores are calculated for all 5′ and 3′ splice junction candidates. In Staden's own words, "all the known junctions give a reasonably high score, but . . . there are many other potential splice sites some of which give even higher peaks." Since no rules for picking acceptable 5′–3′ junction pairs are given, this method appears to be of rather limited value when used alone. Possibly, inclusion of a search for the lariat branch point consensus might to some extent relieve this shortcoming, since the 3′ splice almost invariably involves the first AG found in a rather narrow distance range downstream from the lariat loop.

Taking a more eclectic approach, Nakata *et al.* (1985) include information not only from the 5′ and 3′ consensus patterns, but also from the estimated free energy of snRNA–RNA base pairing and from known statistical differences between the nucleotide sequences of coding and noncoding regions in their prediction scheme. A Perceptron algorithm (see Section V,A,2) is used to generate weight matrices similar to the ones used by Staden. The maximal stabilities of snRNA–RNA base pairings are estimated by a dynamic programming method such as described in Section VII,B. Finally, the likelihood that sequences on both sides of a putative intron–exon boundary are coding or noncoding is estimated by Fickett's method (Section VI,B,2).

Through the use of discriminant analysis—a way to assign optimal weights to a set of parameters characterizing "true" and "false" splice sites such that the resulting aggregate measure maximizes the discrimination between the two sets (cf. Chapter 5, Section IV)—an algorithm is developed to delineate both the 5′ and 3′ ends of the next intron, starting from a given open reading frame. The procedure is as follows: (1) Locate the nearest downstream termination codon. (2) Find all GT and AG pairs preceding the termination codon, and construct a list of all possible GT–AG pairs (i.e., all candidate introns). (3) For each candidate intron, check if the reading frame resulting from splicing continues for at least L_{min} bases after the AG. (4) Retain only those of the remaining candidates that is allocated to the true class by the optimal discriminant function based on the characteristics discussed, namely the scores for the 5′ and 3′ consensus matches, the strength of the putative U1 snRNA base pairing, and the coding/noncoding probability as estimated by Fickett's method.

For $L_{min} = 30$, the authors report that none of the false splice sites in a sample of human sequences from GenBank were mistakenly identified as true, and that 81% of the true sites were correctly found. All in all, 26 true junctions and 60 false ones were included in the analysis. It is not clear to what extent these results depend on the choice of sequences for the training set used in the Perceptron algorithm, however.

This last method may point the way to future integrated prediction schemes where a whole array of measures are considered simultaneously in the prediction of a given signal or functionally distinct section of the DNA. In this sense, it extracts the maximum information from the sequence at hand. From the point of view of having prediction methods that mimic the actual protein–DNA–RNA interactions responsible for a given event, however, it clearly represents a blind alley in that it considers illegitimate information such as the coding/noncoding probabilities that most certainly are never seen by the real splicing machinery.

V. INITIATION OF PROTEIN SYNTHESIS: RIBOSOME-BINDING SITES

Once the mRNA is presented to the protein-synthesizing machinery, specific features in the mRNA must attract ribosomes to the appropriate initiation sites. With the aid of various initiation factors, the initiator Met-tRNA binds to the small ribosomal subunit, then the mRNA binds, and finally the large ribosomal subunit joins the complex: Translation can begin.

Interestingly, although the overall similarity between these steps in prokaryotes and eukaryotes is striking, the patterns used for recognition of the initiator region on the mRNA are rather different. Thus, prokaryotic ribosomes can initiate protein synthesis from internal sites, whereas in eukaryotic mRNAs only the most 5′ site(s) is utilized (Kozak, 1983). As will become clear below, this difference is directly apparent in the patterns of conserved nucleotides around the initiator codons.

A. Ribosome-Binding Sites in Prokaryotic mRNAs

1. Experimental Background

Since prokaryotic ribosomes can find initiation sites in any region of the mRNA, polycistronic messengers are possible and indeed are the rule in the bacterial world. AUG is by far the most frequently used initiator codon, but its close relatives GUG, UUG, and AUU are also found to be used occasionally. Obviously, most AUGs do not define translation start sites—the number of potential open reading frames can be at least an order of magnitude larger than the number of proteins really encoded—and further restrictions on the permissible sequences are long since known, in particular the so-called Shine–Dalgarno sequence (Shine and Dalgarno, 1974). This is a purine-rich stretch of some 5–10 residues centered about 10 bases upstream from the AUG, and it mediates ribosome binding by virtue of its complementarity to a highly conserved region (CUCCU) near the 3′ end of the 16S rRNA (Steitz and Jakes, 1975). Mutations in the Shine–Dalgarno sequence and in the AUG: mutations leading to the AUG being sequestered inside a stable RNA hairpin; changes in the Shine–Dalgarno—AUG distance beyond the optimal 7 ± 2 bases; and mutations in the first ~ 10 bases of the coding region, all impair translation initiation, whereas the region upstream of the Shine–Dalgarno sequence seems to matter less (Kozak, 1983; Tessier *et al.*, 1984).

2. *Theoretical Analyses and Prediction Schemes*

Stormo *et al.* (1982a) have carried out rather extensive statistical studies on a sample of 124 known prokaryotic ribosome-binding sites. The Shine–Dalgarno complementarities to the 16S rRNA are centered around the conserved 5′-UCC-3′ segment of the latter some 6–13 bases upstream from the initiator AUG (Fig. 4.12). The upstream region is generally rich in A residues and lacking in G residues, and a similar trend in the early parts of the coding region may result from a biased third-letter codon choice (Rodier *et al.*, 1982).

After trying a number of rule-based prediction schemes such as: "locate all sequences where AGG, GGA, or GAG are followed by an AUG within 6 to 9 bases," Stormo *et al.* (1982b) opted for a weight matrix method, similar in outline to those already described in Section II,A,2, but with an optimized weight matrix obtained through a Perceptron algorithm. The object of the Perceptron is to build a weight matrix **W** such that the scores P calculated for all true sequences (included in the set S^+) are strictly larger than the scores for all false sequences (in the set S^-). The sequences must be prealigned from some reference point such as the initiator AUG. Starting from a given **W** (all elements equal to 0, say), the algorithm tests the sequences in S^+ and S^- and builds **W** according to the following scheme:

TEST: choose a sequence S from S^+ or S^-
if S belongs to S^+ and $P \geq T$ go to TEST
if S belongs to S^+ and $P < T$ go to ADD
if S belongs to S^- and $P < T$ go to TEST
if S belongs to S^- and $P \geq T$ go to SUBT
ADD: add S to **W**
go to TEST
SUBT: subtract S from **W**
go to TEST

T is a preset threshold, and ADD and SUBT simply change the weight matrix **W** by either adding or subtracting sequence S from it. The algorithm succeeds when all sequences in S^+ have $P \geq T$ and all sequences in S^- have $P < T$ (there is no *a priori* guarantee that such a solution is possible). In the simpler weight matrix schemes discussed previously, **W** was set up from the outset by including all sequences in S^+, forgetting about S^-, and not optimizing with respect to which sequences should be included in the final **W** and which should not.

Assigning the 124 known ribosome-binding sites to S^+ and 167 false sites found by the one of the rule-based schemes to S^-, the Perceptron indeed found a solution, and produced the weight matrix shown in Fig. 4.13. This **W** by definition can discriminate the 124 true sites from the 167 false ones; it is, not unexpectedly, less successful when applied to

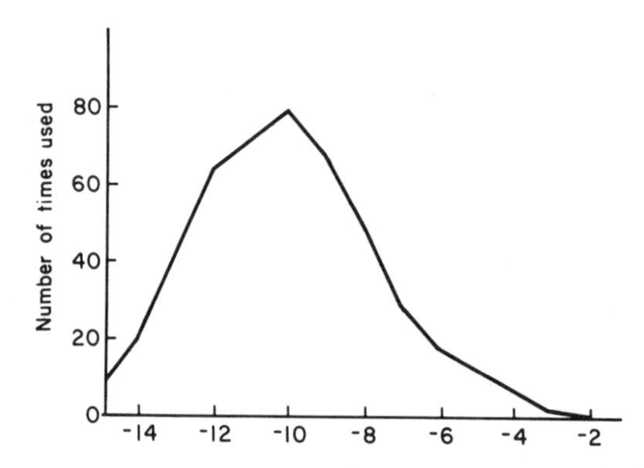

Fig. 4.12. Distribution of the distance between the 5′ base in the initiation codon and the bases that partake in the most stable complementary base-paired structure with 16 S rRNA in a sample of 94 prokaryotic mRNAs (Stormo *et al.*, 1982a).

POSITION	-60	-59	-58	-57	-56	-55	-54	-53	-52	-51	-50	-49	-48	-47	-46	-45	-44	-43	-42	-41	-40	-39	-38	-37	-36	-35
A	7	-2	13	-2	-8	-13	-18	5	0	-5	13	8	-15	9	-4	-7	9	0	-8	-11	-10	-6	-7	-5	-6	-12
C	-21	-6	-11	-21	0	8	-7	-12	-1	1	0	-19	12	-3	-1	10	2	-8	-5	-11	8	1	23	6	-5	2
G	-6	-9	-7	0	8	-16	-4	-2	-16	1	-4	8	-14	5	11	-13	-24	3	7	22	-11	-9	-15	10	-4	4
T	5	1	-3	9	-14	7	15	-5	3	-16	-17	4	18	5	-3	-1	2	4	5	-5	7	8	-5	-15	6	3

POSITION	-34	-33	-32	-31	-30	-29	-28	-27	-26	-25	-24	-23	-22	-21	-20	-19	-18	-17	-16	-15	-14	-13	-12	-11	-10
A	-1	-27	-3	-6	0	-12	-3	-4	-7	14	-2	-4	-6	0	12	5	-9	0	-11	-11	10	8	2	8	0
C	-14	-3	-8	-10	-21	2	0	-2	-1	-11	-3	-1	5	-11	-4	7	0	-14	6	-8	-20	-7	-36	-44	-15
G	-5	-6	-3	-1	-4	-1	-4	-15	0	-14	3	10	-19	-3	-10	-7	-7	7	1	-8	-6	15	21	42	35
T	4	16	-4	7	11	-4	-1	12	8	10	-1	1	8	2	-10	-16	11	1	-3	16	-3	-36	-8	-27	-53

POSITION	-9	-8	-7	-6	-5	-4	-3	-2	-1	0	1	2	3	4	5	6	7	8	9	10	11	12	13	14	15
A	-3	-5	4	-20	-11	5	6	-2	-15	66	-69	-52	-5	-4	6	8	-24	-7	-10	-7	13	14	-9	-18	14
C	-50	-43	-35	-38	-29	-29	1	-9	1	-87	-55	-64	-45	11	-22	-14	-20	-15	-15	-10	-22	-5	2	6	6
G	22	16	-6	-5	-15	-25	-33	-28	-53	-36	-50	107	-5	-37	-44	-27	-15	-23	-16	-29	-47	-17	-29	-15	-23
T	-27	-26	-23	2	-7	-14	-40	-28	0	-53	75	-62	-20	-40	-10	-35	-5	-12	-1	4	14	-23	7	-2	-26

POSITION	16	17	18	19	20	21	22	23	24	25	26	27	28	29	30	31	32	33	34	35	36	37	38	39	40
A	-12	-42	1	-5	-4	-32	12	-10	20	-6	-1	3	-4	4	-10	-1	-2	-14	11	14	-3	2	-13	5	5
C	-8	19	-7	9	-3	17	-2	3	-9	5	22	22	8	-1	1	18	6	11	-10	-8	7	10	0	7	14
G	-7	-1	-6	-17	-4	0	-15	-14	-4	-17	-10	-5	-13	-8	10	-13	-13	9	-4	-3	10	2	4	-8	-21
T	1	4	-7	3	-4	0	-10	8	-18	7	-22	-21	8	4	-3	-6	7	-8	1	-5	-16	-16	7	-6	0

Fig. 4.13. The W101 weight matrix for the region around the initiation codon in 124 prokaryotic mRNAs obtained through the use of a Perceptron algorithm (Stormo *et al.*, 1982b). Note that the weights for the initiation codon (positions 0–2) and the Shine–Dalgarno squence around position −10 are highly biased.

sequences not included in the original S^+ and S^- sets. Out of 10 new genes tested, 7 of the 10 true sites were found, together with 5 false ones.

The main feature lacking in this method, as acknowledged by the authors, is the ability to discriminate against AUGs buried in stable RNA hairpins. This would require consideration not only of the linear sequence, but of dyad symmetries and thermodynamic stabilities of putative hairpins. As will become clear in Section VII, prediction of RNA secondary structure is fraught with difficulties, and analysis of local dyad symmetries may not be sufficient to weed out more than a fraction of all false sites present (Ganoza *et al.*, 1987).

Nevertheless, used in conjunction with methods for locating probable coding regions based on codon usage described below, this method is clearly of great value already in its present form.

B. Ribosome-Binding Sites in Eukaryotic mRNAs

1. Experimental Background

Compared with their prokaryotic *alter egos,* eukaryotic ribosomes seem to have chosen a very different strategy for finding the correct translation initiation sites (see Kozak, 1983, for a review). Rather than roaming around searching for inviting Shine–Dalgarno scents and unashamedly exposed AUGs, they head straight for their target: capped 5′ mRNA ends. The cap—a terminal G methylated in position 7 and linked via a 5′–5′ triphosphate bond to the RNA—both increases the stability of the transcript and enhances the binding of 40S ribosomal subunits to the mRNA. Once bound, the ribosome leisurely wanders off toward the first downstream AUG, where it initiates protein synthesis in some 95% of all mRNAs. Only if the region immediately surrounding this AUG fails to meet some rather liberal criteria does the ribosome bother to continue its search, eventually ending up on a more tasty spot.

This scenario from the modified scanning model (Kozak, 1981) satisfactorily explains many experimental observations such as the requirement for a free 5′ terminus for initiation, and translational arrest upon hybridization to DNA fragments complementary to parts of the 5′ noncoding region, as well as the demonstration that the identities of the two bases occupying positions 3 bases upstream and immediately downstream of the AUG are by and large all that matters for efficient initiation (Kozak, 1984a, 1986a). Also in contrast to what is observed for prokaryotes, only exceptionally stable hairpin loops around the AUG have any effect on protein synthesis in eukaryotes (Kozak, 1986b).

2. Theoretical Analyses

The rudimentary requirements for specific recognition signals implied by the scanning model are readily apparent in the very limited consensus patterns found in eukaryotic mRNAs. A compilation of 211 sequences (Kozak, 1984b) yielded the consensus CC(A,G)CC*AUG*G (initiation codon in italics), with positions −3 (80% A, 18% G) and +4 (40% G) as the only strongly conserved bases. As already noted, these two positions have been shown experimentally to influence the efficiency of initiation (position −3 more so than +4), whereas the C residues only seem to matter when the −3 and +4 bases are suboptimal. A more relaxed consensus, ANN*AUG*, would thus seem to be sufficient in most cases, and indeed, slightly different preferences in the less well-conserved positions become apparent when vertebrate, invertebrate, and plant sequences are analyzed separately (Cavener, 1987; Lütcke *et al.,* 1987).

The function of the conserved bases is still unknown, although attempts to find putative Shine–Dalgarno-like complementarities to regions in the 18S rRNA are legion (see Maroun *et al.,* 1986, for a review; De Wachter, 1979, for a critique). The G in +4 has been suggested to base-pair with the fourth base in the anticodon of the tRNA bearing the initiator Met—again, this has not been tested experimentally.

Even in the absence of long consensus signals, prediction of the most likely start site is exceptionally simple: By always picking the most 5′ AUG, one is assured 90% success! Going beyond this, Kozak (1984b) has suggested continuing the search downstream when the first candidate is embedded in a particularly poor context, especially if the A (or G) in position −3 is absent.

Staden (1984a) has implemented a weight matrix method based on a presumed base-pairing between the mRNA and the 3′ region of 18S rRNA (Sargan *et al.,* 1982); this method does not seem to have been critically evaluated, and rests upon an unproven model for mRNA–rRNA interactions.

VI. FINDING CODING REGIONS

The sheer number of works in this area makes it one of the theoretically most well-studied of all pattern recognition problems in molecular biology. And indeed, as the volume of DNA sequence data grows, reliable methods for picking out likely coding regions in sequences of unknown function will become more and more important. As has already been pointed out, "gene search by signal" routines that look for promoters, terminators, or ribosome-binding sites, are in general less reliable than the "gene search by content" methods described in this section; used in concert, the two approaches provide a rather powerful tool (Staden, 1984b). But before we embark upon our climb toward the upper reaches of theoretical sequence analysis, a brief review of some pertinent aspects of codon usage in natural DNA is needed.

A. Codon Usage

It has long been recognized that synonymous codons—those coding for the same amino acid—are not utilized with equal frequencies, and that different genomes differ in their codon preferences (Grantham, 1978; Grantham *et al.*, 1980; Bennetzen and Hall, 1982; McLachlan *et al.*, 1984; Blake and Hinds, 1984; Maruyama *et al.*, 1986; Sharp *et al.*, 1986). As an example of this, organisms with a high GC content in their genome have a high percentage of G and C in the third-codon position; the other two positions, which are much more constrained by the amino acid sequence, differ much less in GC content between GC-rich and GC-poor organisms (Bernardi and Bernardi, 1985; Wada and Suyama, 1985; Bibb *et al.*, 1984; Aota and Ikemura, 1986). A strong correlation between codon usage and tRNA content is also observed in many organisms (Ikemura, 1981a,b, 1985; Bulmer, 1987).

In addition to this genome constraint on codon usage, contextual constraints that involve a correlation between codons and their immediate surroundings have also been found. Wilbur and Lipman (1983) have shown that a simple genome constraint—codons are strung together randomly with frequencies given by their overall usage—is not sufficient to account for the observed nonrandom dinucleotide frequencies in eukaryotic mRNAs. Yarus and Folley (1985) and Shpaer (1986), analyzing more extensive samples of highly and weakly expressed genes from *E. coli,* come to the conclusion that there is a correlation between the kind of base found in the wobble position (third base) in a codon and the neighboring bases on the 5′ and 3′ sides of the codon; they are not in complete agreement as to whether this effect is stronger for the highly or the weakly expressed genes, however.

The number of explanations advanced for the nonrandom codon usage is almost equal to the number of authors; the suggestions range from maximization of the rate of translation (Pontier, 1970; von Heijne and Blomberg, 1979; Varenne *et al.*, 1984; Liljenström *et al.*, 1985) or of mRNA secondary structure (Nussinov, 1982), to reducing the possibilities for mistranslation (Parker *et al.*, 1983), optimizing the energy of codon–anticodon interaction (Grosjean *et al.*, 1978), or minimizing the energy spent on proofreading the tRNA before its acceptance (Holm, 1986). tRNA–tRNA interactions or the possibility of making a fourth base pair between the anticodon loop and the base immediately 3′ to the codon have been considered likely candidates for being the cause of the codon context effects (Ayer and Yarus, 1986).

No matter what the final consensus on these matters will be—and hotly debated evolutionary issues have a disturbing way of never being settled—the mere existence of strong biases in codon usage has been a great help in the field of theoretical nucleic acid sequence analysis: It can be used to construct fairly reliable methods for finding likely coding regions in newly sequenced DNA.

B. Predicting Coding Regions

Roughly speaking, methods for predicting coding regions or correct reading frames fall in two classes: Those that are based on a previously derived table of codon usage, and those that more generally look for signs of nonrandomness in the DNA sequence. The first kind of method assumes that we know what to expect in terms of amino acid composition and/or codon preferences in the gene(s) we are looking for, and it often quite accurately pinpoints both ends of the coding region, as well as the coding strand and the reading frame used. The second kind requires less *a priori* assumptions and can be applied immediately to DNA from any source without any preliminary codon usage or other tabulation, but it can only be used to estimate the probability that a given segment, defined by the user, is either coding or noncoding. It does not tell which of the two strands is coding, or what the reading frame is, though this can often be quite easily determined through a search for transcriptional or translational signals.

1. Methods Based on Codon Usage

Basically, these methods scan all reading frames of the sequence under study with a "moving window" some 15–25 codons long, and measure, for each position of the window, the degree of similarity between the codons found in the window and the overall codon usage in a relevant standard sample, such as a collection of known genes from the same organism. True coding regions are assumed not to differ substantially in codon usage from the standard, whereas noncoding regions or frames will display a more random usage of codons, and hence have a lower degree of similarity. Various implementations of this idea differ in details of how the similarity is calculated, and in how the standard codon usage table is constructed.

One simple scheme, described by Staden (1984c), takes the expected amino acid composition of the protein searched for as the standard—this might be known from experiments, or one can use the overall composition in some large sample of, say, soluble proteins. Next, a codon usage table is derived on the assumption that all codons corresponding to the same amino acid are equally frequent (from this table, the expected frequencies of T, C, A, and G in the three codon positions can be calculated. They turn out to be quite different from random; thus a definite pattern of base frequencies in the first-, second-, and third-letter positions is expected already from the amino acid composition of a typical protein and the genetic code). Using this table, one can calculate a coding probability, p, for a given DNA segment (typically 25 residues long) and a given reading frame by multiplying together the codon frequencies read off from the table for the codons in the segment. Finally, when this has been done for all three frames, one calculates $P_i = p_i/(p_1 + p_2 + p_3)$, plots $\log[P_i/(1-P_i)]$ for frames $i=$ 1,2,3, moves one codon to the right, and repeats the process. Likely reading frames are often readily seen on the resulting plots, and a simultaneous display of start and stop codons makes their identification even easier (Fig. 4.14).

Obviously, this same scheme can be used with a real codon usage table as found for the kind of organism or the kind of genes one is working with; this will add any bias in codon usage resulting from

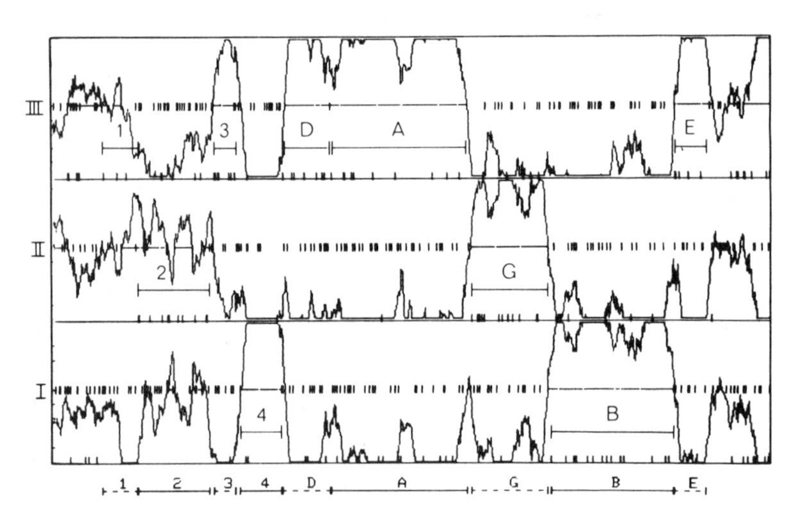

Fig. 4.14. P_1, P_2, and P_3 profiles for the *E. coli unc* operon obtained using a codon table with average amino acid composition and no codon preference. Stop codons are shown as short vertical bars at the 50% level. (From Staden, 1984c.)

selection working on the DNA or RNA levels to the bias caused by the amino acid composition (Staden and McLachlan, 1982; see also Blake and Hinds, 1984, who run the probability profile through a least-squares smoothing routine before it is displayed). Considerations of factors beyond the amino acid composition will enhance the predictive power in some circumstances, in particular when some regions code for proteins of an unusual composition such as membrane proteins; on the other hand, one needs a larger data base for the construction of the codon usage table.

A slightly different way of setting up the codon usage table has been presented by Gribskov *et al.* (1984). These authors neutralize the effect of the overall amino acid composition by calculating a preference parameter for each codon in the following way: First, the frequency of each codon f_{abc} and of each family of isoaccepting codons F_{abc} are found from the data base; then, the quotients f_{abc}/F_{abc} (the frequency of each codon within its own family) are normalized by the same quotients expected for a random sequence of the same base composition, r_{abc}/R_{abc}, where $r_{abc} = f_a f_b f_c$. Thus, $p = (f_{abc}/F_{abc})/(r_{abc}/R_{abc})$, and the codon preference statistic P for a given DNA segment and reading frame is defined as

$$P = p_1 p_2 \ . \ . \ . \ p_L \tag{4.3}$$

where L is the window length. The authors recommend $L = 25$ for sequences less than 5 kb long, and $L = 50$ for longer sequences. They also suggest using the codon preference parameters for a set of highly expressed genes of the organism in question as the standard of choice.

In addition to its use for finding likely coding regions or sequencing errors, the mean P value calculated over a whole gene provides a good indication of the level of expression *in vivo,* since it is independent of the amino acid composition of the corresponding protein and only depends on the degree of match to the optimal composition of isoaccepting codons. In this sense, it is superior to the method of Staden. A similar measure, the "Codon Adaptation Index," has been proposed by Sharp and Li (1987).

Instead of using the full codon usage table as the standard against which the sequence is tested, one can focus on the expected positional base frequencies in codon positions one, two, and three (Staden, 1984c). These can be calculated either from the expected overall amino acid composition assuming equal codon usage within each codon family, or from the full codon usage table.

Starting from a table of the positional base frequencies $E(b,i)$ of the standard (with $b = $ A,T,G,C and $i = 1,2,3$), for each position of the window over the test sequence one calculates the sum s of the corresponding Es taken from the table. This is repeated for all three reading frames, and the relative scores $S_i = s_i/(s_1 + s_2 + s_3)$ are plotted as functions of the window position. Not surprisingly, this method has a higher noise level than those based on the full codon usage table—going from 20 amino acid frequencies or 64 codon frequencies to 12 positional base frequencies amounts to a good deal of data reduction—but may be a better choice when the data base used in the construction of the codon usage table is small.

Shepherd's (1981a) method was an early attempt along the same lines, based on what was thought to be a remnant of a primitive genetic code of the form RNY (R = purine, Y = pyrimidine, N = any base, Shepherd, 1981b; but see also Staden, 1984c). For any given position of the window, the deviation of each of the three frames from the assumed RNY pattern is measured by calculating the smallest number of R→Y and Y→R mutations required to make all codons in the window RNY. If one frame has a consistently better fit to the RNY pattern over a long stretch of successive window positions, this is taken as an indication of a likely coding region. According to Staden (1984c), the positional base frequency method is superior to the RNY method—again not very surprisingly, since the RNY pattern contains even less information than the positional base frequencies.

2. *Methods Based on Local Nonrandomness in the Sequence*

The most well-known of these methods is, no doubt, Fickett's (1982) TESTCODE program. Its aim is to assign a given region of the sequence, defined by the user at the outset, to one of three classes: coding, noncoding, or no opinion. Once the region to be analyzed has been given to the program, the classification is fully automatic with no elements of subjective evaluation by the user. This also allows the reliability of the method to be quantified (in marked contrast to most of the methods described in the previous section): in its published form, TESTCODE returns the answer "no opinion" on some 18% of all sequences; in addition, it incorrectly labels some 6% of all coding sequences as noncoding; and it labels some 3% of all noncoding sequences as coding.

The method is based on the calculation of a weighted sum of eight parameters determined for the sequence under scrutiny. The first four measure the asymmetry in the distribution of each base among the three codon positions. With

$$\begin{aligned} A_1 &= \text{number of As in positions } 1,4,7,10, \ldots \\ A_2 &= \text{number of As in positions } 2,5,8,11, \ldots \\ A_3 &= \text{number of As in positions } 3,6,9,12, \ldots \end{aligned} \tag{4.4}$$

one can define the A asymmetry by

$$A_{as} = \text{Max}(A_1,A_2,A_3)/[\text{Min}(A_1,A_2,A_3) + 1] \tag{4.5}$$

and similarly for G_{as}, C_{as}, and T_{as}. The other four parameters are simply the frequencies of the four bases in the sequence: f_A, f_G, f_T, and f_C.

The distribution of these eight parameter values in a sample of 321 coding and 249 noncoding sequences, all longer than 200 bp, are given in Table 4.1. From Table 4.1, we see that, e.g., 94% of all sequences with $A_{as} > 1.9$ are coding, and that only 28% of all sequences with f_A in

TABLE 4.1. **Probability-of-Coding Values Used in Fickett's (1982) Method**

	Probability of coding			
Parameter	A	C	G	T
Asymmetry				
0.0 to 1.1	0.22	0.23	0.08	0.09
1.1 to 1.2	0.20	0.30	0.08	0.09
1.2 to 1.3	0.34	0.33	0.16	0.20
1.3 to 1.4	0.45	0.51	0.27	0.54
1.4 to 1.5	0.68	0.48	0.48	0.44
1.5 to 1.6	0.58	0.66	0.53	0.69
1.6 to 1.7	0.93	0.81	0.64	0.68
1.7 to 1.8	0.84	0.70	0.74	0.91
1.8 to 1.9	0.68	0.70	0.88	0.97
1.9 to 2.0+	0.94	0.80	0.90	0.97
Content				
0.00 to 0.17	0.21	0.31	0.29	0.58
0.17 to 0.19	0.81	0.39	0.33	0.51
0.19 to 0.21	0.65	0.44	0.41	0.69
0.21 to 0.23	0.67	0.43	0.41	0.56
0.23 to 0.25	0.49	0.59	0.73	0.75
0.25 to 0.27	0.62	0.59	0.64	0.55
0.27 to 0.29	0.55	0.64	0.64	0.40
0.29 to 0.31	0.44	0.51	0.47	0.39
0.31 to 0.33	0.49	0.64	0.54	0.24
0.33 to 0.99	0.28	0.82	0.40	0.28

the interval 0.33 to 0.99 are coding. By assigning appropriate weights $w_1 \ldots w_8$ to the parameters (Table 4.2), a unique score can be calculated for any sequence as follows: First, calculate the eight parameter values for sequence; then, look up the corresponding probability of coding values p_i in Table 4.1; and finally, calculate the sum $S = \Sigma w_i p_i$, $i = 1,..,8$. Depending on the value of S, the prediction is that the sequence is noncoding ($S < 0.74$), no opinion ($0.74 < S < 0.95$), or coding ($S > 0.95$).

Clearly, this algorithm cannot predict the correct frame, or even the correct strand. It will, however, pick up any pattern of nonrandom codon usage such that the different bases have different frequencies in the three codon positions; positional preferences that do not have a period that is a multiple of three (such as dinucleotide preferences) will not be picked up. Thus, it is much more general than the methods based explicitly on codon usage, but its resolution is poorer, since it needs a region at least 200 bp long to give reliable results. The exact boundaries of the putative coding region being analyzed must be determined by other means, such as some gene search by signal method.

The Uneven Positional Base Frequencies algorithm of Staden (1984c) is very similar to Fickett's method, but the measure of nonrandomness is slightly different. Starting from the A_1, A_2, and A_3 values as defined above,

$$\begin{aligned} \text{AMEAN} &= (A_1 + A_2 + A_3)/3 \\ \text{ADIF} &= \text{abs}(A_1 - \text{AMEAN}) + \text{abs}\,(A_2 - \text{AMEAN}) + \\ &\quad \text{abs}(A_3 - \text{AMEAN}) \end{aligned} \tag{4.6}$$

are calculated, together with the corresponding GDIF, CDIF, and

TABLE 4.2. **Weights Used for Calculating *S* Values in Fickett's (1982) Method**

Residue	Asymmetry	Content
A	0.26	0.11
C	0.18	0.12
G	0.31	0.15
T	0.33	0.14

TDIF values—these values are independent of the overall base composition. The prediction is then made on the basis of the value of

$$S = \text{ADIF} + \text{GDIF} + \text{CDIF} + \text{TDIF} \tag{4.7}$$

but the actual cutoff used is not given in the reference.

A third method of the same kind and with similar predictive powers, but couched in the rather unyielding mathematical language of Principal Component Analysis has recently been added to the list (Michel, 1986).

3. *Other Methods*

A number of other, less well known, methods that in one way or another exploit preferences for certain nucleotides to appear in certain codon positions are also to be found in the literature. Thus, Shulman *et al.* (1981) was one of the pioneers in the field with a method where the correct reading frame is identified by virtue of the fact that there is a reasonably strong correlation between the kind of nucleotide found in position 1 and the kind found in position 2 of codons, as well as between positions 2 and 3, whereas the correlation between position 3 and position 1 of the following codon is much weaker. By calculating the magnitudes of these pairwise correlations in a putative coding region, the correct frame can be found.

The third codon position being less restricted by the amino acid sequence than the other two leads to the hypothesis that this position should be free to vary in response to DNA- or RNA-level constraints. In some mitochondrial genes, this is apparent as a highly skewed distribution of nucleotides in the third codon position that can be used to infer the correct reading frame (Almagor, 1985; this does not seem to generally true, however). Similarly, in genomes of either very high or very low GC content, the third codon position is much more biased toward or away from GC than the other two, and a simple plot of the GC content of the three positions will give clues as to which is the correct reading frame (Bibb *et al.,* 1984).

One can also choose to focus on the amino acid sequence rather than on the codons: Dean *et al.* (1984) compare the actual amino acid composition of the sequence obtained by translating the putative gene with that expected from a random nucleotide sequence of the same overall base composition. If the two differ substantially (as judged by χ^2 analysis), the assumption is that we are dealing with a true coding region.

To summarize: "gene search by content" rather than "gene search by signal" methods are the first to be tried when searching for unknown coding regions. If a sufficient amount of sequence data is already available for the genome in question, methods that compare the codon usage in the new sequence with the genome standard have a higher resolution and indicate both the coding strand and the reading frame. If no reliable codon usage table can be constructed, methods that measure local nonrandomness are still available. Both methods can be combined profitably with a search for transcription and translation consensus signals, but the apparent absence of such signals around a region that has been given a high coding score should not be taken to imply necessarily that it cannot be transcribed and translated. At this point, it is not possible to tell which of the methods described in this section is the best; as with all kinds of computer analysis, one should entertain a healthy skepticism whenever seated in front of a keyboard and monitor.

VII. RNA FOLDING AND STRUCTURE

So far, we have focused on patterns that can be found in the primary sequence of DNA or RNA; we now turn to higher-order structures. The forces guiding the folding of an RNA molecule, be it a tRNA, an rRNA, or an mRNA, is not particularly well understood—in part because the folded structure is rather hard to probe biochemically, in part because most RNAs do not crystallize and hence cannot be analyzed by X-ray diffraction. Nevertheless, to the extent that standard Watson–Crick base-pairing determines the final structure(s) and to the extent that only one or a few particularly stable structures represent the final state, the full thermodynamic description of the folding energetics can be approximated by a combinatorial approach where the maximally base-paired or energetically most favored combination of base pairs is sought. Formulated in this way, the folding problem is much more tractable theoretically. Algorithms that are guaranteed to find the lowest energy structure, given a sequence and a set of experimentally derived energy terms for base pairs and loops, are known since a number of years and can currently analyze sequences up to 1–2 kb long (Fig. 4.15). Unfortunately, one cannot trust the output of these programs to even approximately represent the true *in vivo* structure.

The present state-of-the-art can perhaps best be appreciated when it

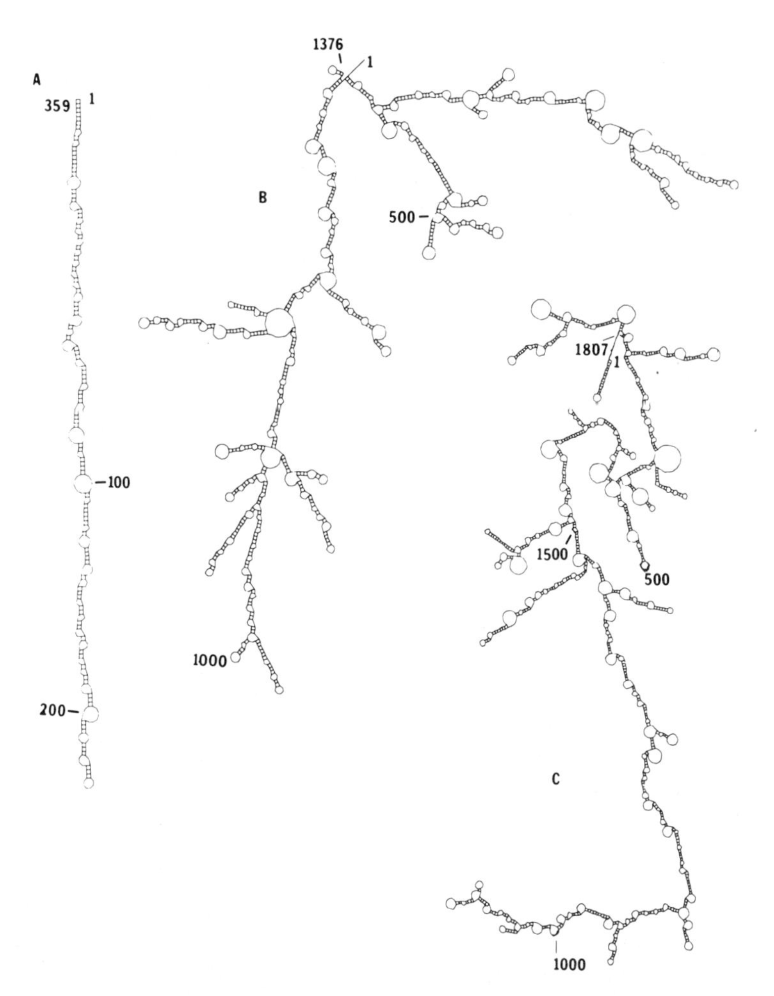

Fig. 4.15. Predicted minimal energy structures for the 359-bp circular RNA of potato spindle tuber viroid (A), and the *E. coli* 1376-bp *ompA* (B) and 1807-bp *ompF* (C) transcripts. (From Mount, 1984.)

is recognized that an RNA molecule can start to fold in a 5′→3′ direction as it is being synthesized, i.e., there may be a decisive kinetic element that in part determines which complementary regions will in fact pair and which will not. Further, the initial folded structure may change its conformation by local (or global?) rearrangements of the base pairs, thus relaxing toward lower energy states that may be either local or global minima. Third, tertiary interactions between unpaired base pairs in loops or at the ends of the molecule, and additional reshufflings of already formed base pairs to accommodate these new interactions, may alter the structure. And finally, most RNAs are strongly bound to other RNAs or to proteins when in their functional state. The secondary structure prediction methods currently in use are far from capturing all the essentials of this complex folding process, but many of them allow the user to enter independent experimental information on known single- or double-stranded sections of the chain, thus constraining the program to explore only a narrow region around the true structure.

In this section, published methods will be dealt with according to their basic approach to the problem. Dot matrix methods, being conceptually the most simple ones, will be described first. Global minimum energy search methods are next, followed by stem-and-loop analysis, i.e., methods that start from a list of all locally stable hairpin loops. Finally, a couple of interactive methods, an approach based on homologous sequences, and two algorithms designed to find tRNA genes are presented.

A. Dot Matrix Methods

Graphical matrices have the obvious advantage that all possible self-complementary pairings are simultaneously presented to the eye —also, these methods are easily programmed even on a microcomputer and can analyze many kilobases of sequence in a rather short time. Normally, the sequence is written in the 5′→3′ direction both across and down the side of a matrix, and dots are placed in all cells of the matrix where the two bases as read off from the horizontal and vertical sequences are complementary, i.e., when they can form a G-C, A-U, or G-U base pair (Fig. 4.16). Self-complementary regions show up as pairs of short diagonal lines perpendicular to and symmetrically placed *vis à vis* the main upper left to lower right diagonal. Bulge loops, internal loops, and bifurcation loops all have their own special fingerprints on the matrix plot (Fig. 4.17).

Since about one-third of the matrix will be filled with dots even for a random sequence, some kind of filtering is usually applied before the results are displayed. The simplest filter checks segments of n bases at a time and only prints a dot when at least m out of n bases are complementary—this efficiently suppresses a lot of random noise.

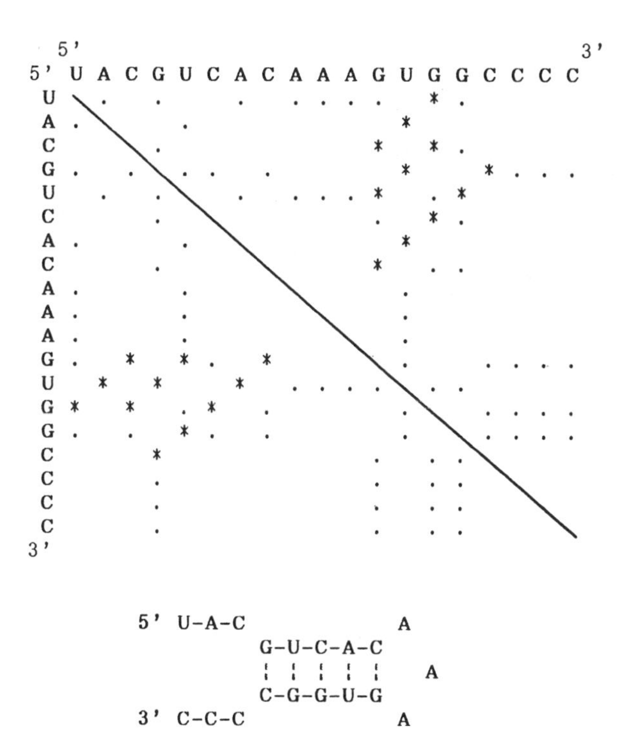

Fig. 4.16. Dot matrix showing self-complementary regions in a short RNA. Regions with three or more out of five consecutive matches are shown by *. The "best" structure is also shown.

Color coding can be used to display matrices obtained with different stringency on the same plot. A typical low-stringency search would use $m = 3, n = 5$ (marked by * in Fig. 4.16); a high-stringency search might use $m = 9, n = 11$ (Maizel and Lenk, 1981).

A good example of more advanced filtering techniques is provided by the method of Quigley *et al.* (1984). The first filter only accepts segments with a minimum number of contiguous complementary bases (4, say), and further requires that the two complementary segments are at least 3 bases apart (i.e., a strip 3 bases wide along the main diagonal is excluded from consideration). A second filter calculates the base pairing energies of the regions passed on from filter 1, and rejects all segments with energies above a certain threshold. Finally, a third filter checks the segments passed on from filter 2 against any experi-

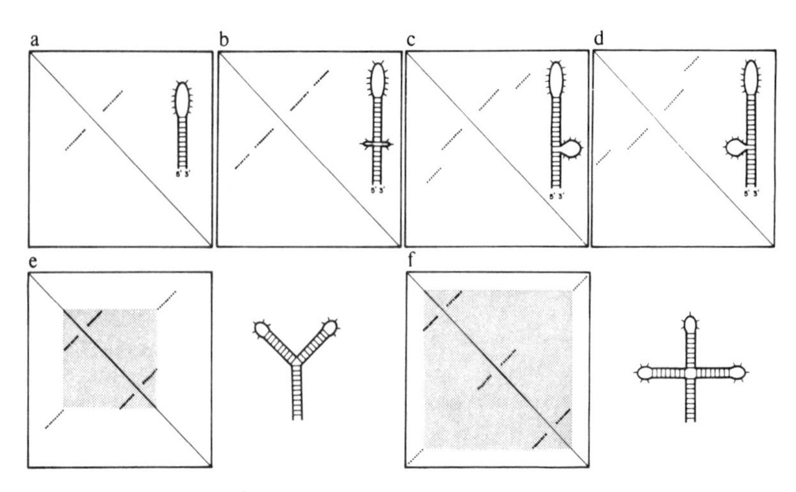

Fig. 4.17. Some simple stem-and-loop structures as they appear in a dot matrix. (From Quigley *et al.*, 1984.)

mental information available (single- or double-strand-specific nuclease digestion, chemical cross-linking, etc.), and the user can in this way instruct the program to suppress all structures not compatible with known or suspected single- and double-stranded regions. Even after all these filters have been applied, quite a number of mutually inclusive or exclusive pairings usually remain, and a whole set of equally likely structures can be constructed (see Quigley *et al.*, 1984, for some useful hints helping in the interpretation of the matrix plot). This is more of an advantage than a disadvantage at our present level of ignorance, since a range of possible structures may suggest experimental tests or functional possibilities that a single best (probably erroneous!) structure would not.

Dot matrix methods are thus within the reach of even the simplest PC, are fast, can be run under different stringencies thus focusing attention first on the strongest, then on successively weaker complementarities, and allow the experienced user to explore many alternative structures in a reasonable time. In short, they are subjective in the good sense of the word.

B. Global Minimum Energy Structures

To predict the most likely secondary structure(s) of an RNA sequence, one could imagine a brute force approach where the computer systematically generates all conceivable structures and evaluates their energies according to some thermodynamic rules. This would not only give the energetically most favorable base-pairing scheme, it would also provide us with all other structures and their energies, thus allowing the complete statistical mechanics of the molecule to be worked out: The probability of finding the molecule in any given structural state at any given temperature could then easily be calculated.

Unfortunately, combinatorial problems of this kind have a bad habit of quickly outgrowing even the fastest supercomputer. A gross estimate of the number of different structures that an RNA chain of length N can adopt leads to the formula (Zuker, 1986):

$$T(N) = cN^{-3/2}(2.29)^N \tag{4.8}$$

where c is between 0.1 and 10. For $N = 120$ bases, this amounts to some 10^{40} different structures—obviously beyond the capabilities of any machine imaginable.

If we settle for something less than the full enumeration of all possible structures, however, the situation improves. One possibility is to search only for the lowest energy structure, to the exclusion of all other possible ones, even those that are of no more than infinitesimally higher energy (or even of the same energy, if the lowest energy state is degenerate). This problem indeed can be solved by algorithms with computation times that grow roughly as N^3 rather than exponentially. Here, I will describe in some detail a simple scheme that finds the maximally base-paired rather than the lowest energy structure; more elaborate schemes that explicitly include the different base-stacking and loop energies are based on the same principles, and will be presented only in outline.

The algorithm of Nussinov and Jacobson (1980) was the first published dynamic programming algorithm for predicting the secondary structure of RNA molecules, though similar algorithms had long been used in the construction of optimal alignments between related DNA and protein sequences (Chapter 6). To understand the method, it is easier to picture the RNA in a circular representation rather than in the more familiar stem-and-loop form (Fig. 4.18). Base pairs are drawn as arcs connecting the bases, and, since knotted structures are assumed not to be present, all arcs in an acceptable structure must be nonintersecting.

The basic idea is first to determine the optimal structures for short subsections of the chain, and then to use these results when considering

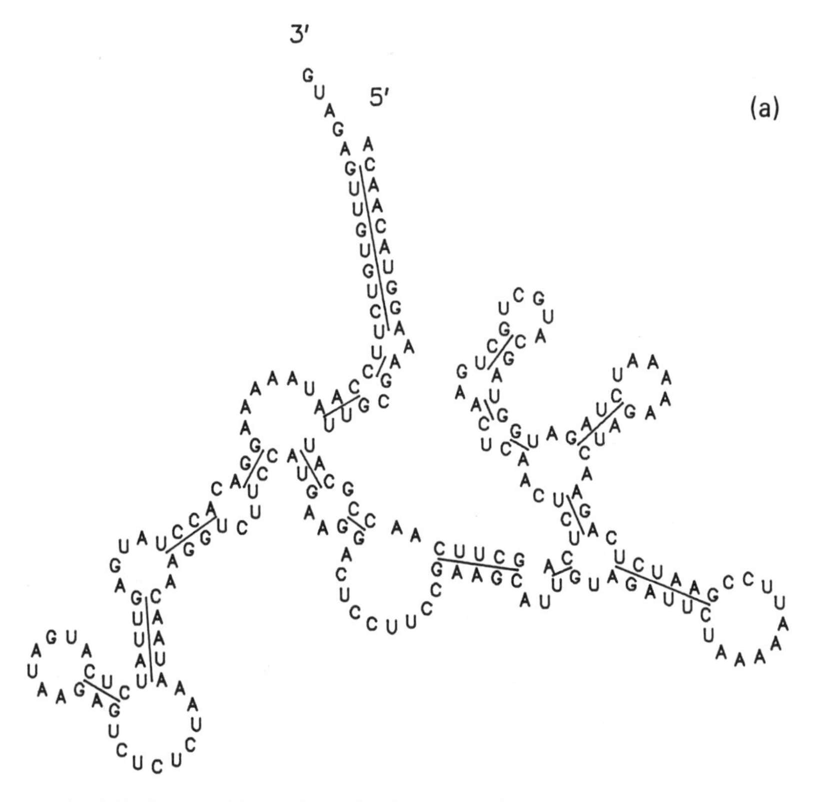

Fig. 4.18. Stem-and-loop (a) and circular (b) (Nussinov circle) representations of the secondary structure of an RNA molecule. The 5′ and 3′ ends correspond to nucleotides number 750 and 955. (From Zuker, 1986, with permission from the National Research Council of Canada.)

progressively longer sections. Consider a subsection $B_i \ldots B_j$ from the sequence $B_1 \ldots B_n$. Now, the ability of base B_k to pair with B_j is tested for $k = i, \ldots, j-1$; if B_k and B_j can pair, the algorithm checks the maximum number of base pairs present in sections $B_i \ldots B_{k-1}$ and $B_{k+1} \ldots B_{j-1}$. After all positions k in the range $i, \ldots, j-1$ have been tested in this way, the maximum number of base pairs found for any k and the position of the base paired with B_j in this case are recorded as elements $M(i,j)$ and $K(i,j)$ of the base pair and traceback matrices **M** and **K**. If B_j cannot pair with any base in $B_i \ldots B_{j-1}$, then one puts $M(i,j) = M(i,j-1)$. The maximum number of base pairs for sections of length p are thus found

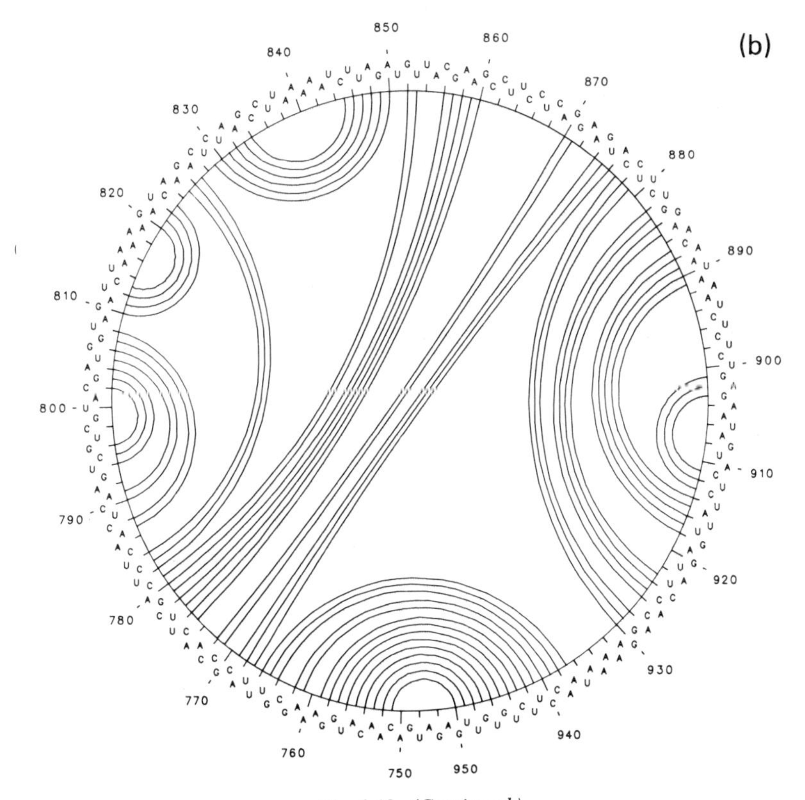

Fig. 4.18 (*Continued*)

by induction from the results already obtained for lengths $p-1$, $p-2,..,2$:

$$M(i,j) = \max\begin{cases} M(i,k-1)+M(k+1,j-1)+1 & [B_k \text{ and } B_j \text{ can pair}] \\ M(\mathrm{i},\mathrm{j}-1) & [\text{no } B_k \text{ can pair with } B_j] \end{cases} \quad (4.9)$$

where $k = i,..,j-1$. Clearly, $M(i,j) = 0$ and $K(i,j) = 0$ for all $i \geq j$, i.e., below and including the main diagonal.

When the two half-matrices **M** and **K** have been filled, the maximum number of base pairs possible is stored in element $M(1,n)$. The corresponding element $K(1,n)$ gives the position of the base B_p which is paired with B_n in the optimal structure; this divides the sequence into subsequences $B_1 \ldots B_{p-1}$ and $B_{p+1} \ldots B_{n-1}$ and the positions of those bases that are paired with B_{p-1} and B_{n-1} in the optimal structure can be read off from $K(1,p-1)$ and $K(p+1,n-1)$, and so on.

As an example, the **M** and **K** half-matrices for the sequence analyzed by the matrix method above (Fig. 4.16) are shown in Fig. 4.19—note that the **M** matrix has been written below the diagonal such that $M(i,j)$ refers to the element in column i and row j. The first diagonal next to

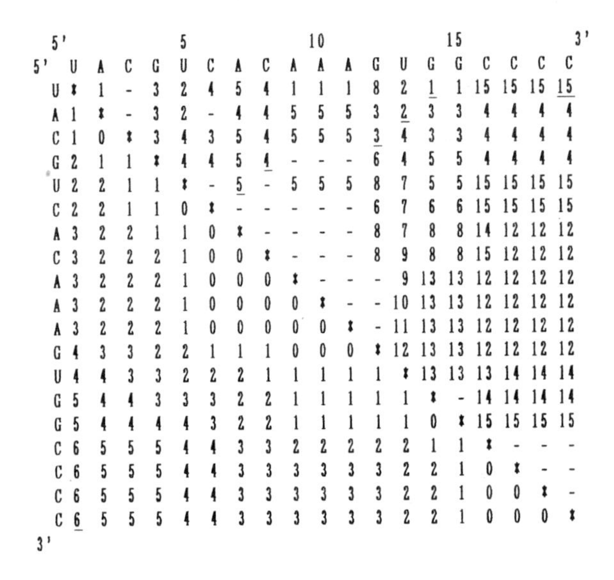

	5'				5					10					15				3'
5'	U	A	C	G	U	C	A	C	A	A	A	G	U	G	G	C	C	C	C
U	*	1	-	3	2	4	5	4	1	1	1	8	2	1	1	15	15	15	15
A	1	*	-	3	2	-	4	4	5	5	5	3	2	3	3	4	4	4	4
C	1	0	*	3	4	3	5	4	5	5	5	3	4	3	3	4	4	4	4
G	2	1	1	*	4	4	5	4	-	-	-	6	4	5	5	4	4	4	4
U	2	2	1	1	*	-	5	-	5	5	5	8	7	5	5	15	15	15	15
C	2	2	1	1	0	*	-	-	-	-	-	6	7	6	6	15	15	15	15
A	3	2	2	1	1	0	*	-	-	-	-	8	7	8	8	14	12	12	12
C	3	2	2	2	1	0	0	*	-	-	-	8	9	8	8	15	12	12	12
A	3	2	2	2	1	0	0	0	*	-	-	-	9	13	13	12	12	12	12
A	3	2	2	2	1	0	0	0	0	*	-	-	10	13	13	12	12	12	12
A	3	2	2	2	1	0	0	0	0	0	*	-	11	13	13	12	12	12	12
G	4	3	3	2	2	1	1	1	0	0	0	*	12	13	13	12	12	12	12
U	4	4	3	3	2	2	2	1	1	1	1	1	*	13	13	13	14	14	14
G	5	4	4	3	3	3	2	2	1	1	1	1	1	*	-	14	14	14	14
G	5	4	4	4	4	3	2	2	1	1	1	1	1	0	*	15	15	15	15
C	6	5	5	5	4	4	3	3	2	2	2	2	2	1	1	*	-	-	-
C	6	5	5	5	4	4	3	3	3	3	3	3	2	2	1	0	*	-	-
C	6	5	5	5	4	4	3	3	3	3	3	3	2	2	1	0	0	*	-
C	6	5	5	5	4	4	3	3	3	3	3	3	2	2	1	0	0	0	*

3'

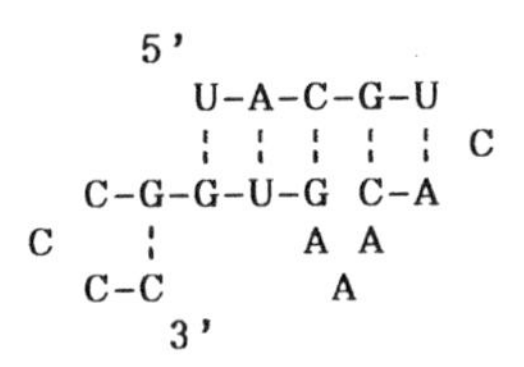

Fig. 4.19. **M** and **K** half-matrices for the same RNA sequence as in Fig. 4.16. Note that the **M** matrix has been written below the diagonal, exchanging the i and j indices. The traceback through **K** is indicated by the underlined elements, and the resulting, maximally base-paired structure is shown at the bottom.

the main one in matrix **M** simply shows which nearest neighbors can pair, and the corresponding elements in the **K** matrix give the position of the 5′ base in each pair; the second diagonal shows which subsequences of length $p = 3$ can form an internal base pair (notice that the choice of which of the two bases that pairs with the 3′ one can be ambiguous); etc. For any given element $M(i,j)$, all previous entries to its right in the row above, $M(k + 1, j - 1)\, k = 1, .., j - 1$, and all entries above it in its own column, $M(i,k - 1)$, are considered when the maximum number of base pairs for the subsection $i, \ldots, j$ is derived according to the formula above.

The maximum number of base pairs in our example is $M(1,19) = 6$, and the positions of these are read off in succession from elements $K(1,19) = 15$ (i.e., a base pair between bases 15–19); $K(16,18) = 0$ (no base pair); $K(1,14) = 1$ (base pair, 1–14); $K(2,13) = 2$ (base pair, 2–13); $K(3,12) = 3$ (base pair, 3–12); $K(4,11) = 0$, $K(4,10) = 0$, $K(4,9) = 0$, $K(4,8) = 4$ (i.e., the optimal base pairs in subsection 4–11 are the same as in 4–8, and 5–11 remain unpaired); $K(5,7) = 5$ (base pair, 5–7).

The optimal structure derived in this way is different from the structure found by the matrix method, though both have the same number of base pairs (note that a pair can be formed between bases 1 and 2 in Fig. 4.16). This illustrates a fundamental shortcoming of the optimal solution method: If degenerate solutions exist, only one will be found (which one depends on how one chooses between degenerate possibilities for the subsegments).

Obviously, the solution found in this case, although optimal from the point of view of the number of base pairs, is a poor structure in terms of the known energetics of hairpin loop formation. As currently understood, hairpin loops are stabilized by stacking interactions between adjacent base pairs in the stem, and are destabilized by the unpaired loops connecting the stems. The most commonly used table of energies is by Salser (1977) (Table 4.3). [See Freier *et al.* (1986) for more recent experimental data.] From these energy terms, the total energy of any structure can be calculated. Algorithms that find the optimal structure of lowest energy rather than the maximum number of base pairs can be constructed following the same idea as described above, only now one calculates the minimum *energies* for all subsegments $i, \ldots, j$, which is slightly more complicated. The traceback procedure, once the **M** and **K** matrices have been filled, is the same. A lot of effort has been put into making these algorithms faster and less mem-

TABLE 4.3. Stacking and Loop Energies[a] as Compiled by Salser (1977) and Tabulated by Zuker (1986)

Stacking energies (UG = GU)

Interior closing pair	Exterior closing pair: GU	AU	UA	CG	GC
GU	−3	−3	−3	−13	−13
AU	−3	−12	−18	−21	−21
UA	−3	−18	−12	−21	−21
CG	−13	−21	−21	−48	−43
GC	−13	−21	−21	−30	−48

Bulge loop destabilizing energies by size of loop

	1	2	3	4	5	6	7	8	9	10	12	14	16	18	20	25	30
	28	39	45	50	52	53	55	56	57	58	59	61	62	63	64	65	67

Hairpin loop destabilizing energies by size of loop

	1	2	3	4	5	6	7	8	9	10	12	14	16	18	20	25	30
CG closing	999	999	84	59	41	43	45	46	48	49	50	52	53	54	55	57	59
AU closing	999	999	80	75	69	64	66	68	69	70	71	73	74	75	76	77	79

Interior loop destabilizing energies by size of loop

Closed by	1	2	3	4	5	6	7	8	9	10	12	14	16	18	20	25	30
CG-CG	999	1	9	16	21	25	26	27	28	29	31	32	33	34	35	37	39
CG-AU	999	10	18	25	30	34	35	36	37	38	39	40	41	42	43	45	47
AU-AU	999	18	26	33	38	42	43	44	45	46	48	49	50	51	52	54	56

[a]Measured in tenths of a kilocalorie per mole.

ory-consuming (for reviews, see Mount, 1984; Zuker and Sankoff, 1984; Zuker, 1986), and using modern supercomputers it is now possible to find the lowest energy structure for sequences up to around 2000 bp long in less than an hour of CPU time.

A useful feature available in most of the more recent programs is the possibility to incorporate information on regions known from experiment to be either base-paired or single stranded; this makes it more probable that the structure found by the computer is reasonable close to the *in vivo* form (Zuker and Stiegler, 1981). An example of this approach can be found in the β-globin mRNA model proposed by Lockard *et al.* (1986).

Although an important aspect of the RNA folding problem—finding the lowest energy structure—is thus solved in a mathematical sense, large uncertainties regarding the biological significance of the resulting structures remain (Zuker, 1986). Thus, most tRNAs are not folded into the well-known cloverleaf form by these algorithms, and one can often find very different structures of almost the same total energy as the optimal one. In this sense, the problem is ill-conditioned: Small changes in the input (energy) parameters can result in very different solutions. It is possible to alter the algorithm such that all solutions that differ from the optimal one by no more than a given energy increment are found (Zuker, 1986; Williams and Tinoco, 1986), but there is still a difficult tradeoff between a too-narrow energy increment and too many solutions. Also, higher-order interactions such as between two unpaired loops (known to be important for tRNAs) or between RNA and protein, being formally excluded from the minimum energy algorithms, presumably can alter the final structure to an unknown extent.

C. Stem-and-Loop Analysis

The third major approach to predicting the secondary structures of RNA molecules is first to locate all possible base-paired stems one by one, and then use some heuristic rules for building structures from these primitives. Finding the stems is straightforward in principle—it amounts to finding all dyad symmetries (or inverted repeats) in the sequence, and this can be accomplished, e.g., by an adapted version of one of the usual homology-search programs (Kanehisa and Goad, 1982; Martinez, 1983; see Chapter 6).

An immediate use of these programs is to calculate the base-pairing potential (i.e., the number of dyad symmetries) of various natural and random sequences; interestingly, it seems that most natural RNAs can form approximately the same number of base pairs as random sequences of similar composition. This does not necessarily mean that natural RNAs are "random" from the point of view of secondary structure: Specific motifs may be selected in specific positions (such as around sites of translation initiation or in the cloverleaf structure of tRNAs), even though the overall base-pairing capabilities are close to the random expectation (Fitch, 1974, 1983; Kanehisa and Goad, 1982).

Programs by Pipas and McMahon (1975) and Studnicka *et al.* (1978) are early examples of algorithms that build acceptable structures from a list of all possible stems; here, I will describe the method of Dumas and Ninio (1982), as revised by Papanicolaou *et al.* (1984), to give the flavor of these schemes.

After a first pass through the sequence during which all possible stems are located (taking GU-pairing into account, and only saving stems with a sufficiently low calculated free energy), this list of segments is partitioned into subsets, or incompatibility islets, where each segment of an islet is incompatible—i.e., cannot coexist—with any of the other segments in the islet. The tree of possible structures is then built by always adding the most stable segment compatible with the growing solution from the set of islets that have not yet contributed a segment to the structure. When multiple solutions of low energy are sought, the energy of a growing solution, plus the sum of the lowest energy stems in all remaining islets, can be compared with the energies of the "best" solutions already found, and that particular branch can be dropped from further consideration if its estimated minimum free energy is too large compared to the solutions already found. With a set of energy values that was originally optimized to predict tRNA structures, this program found the correct fold in approximately 80% of 200 tRNA sequences, and in 60% of 100 5S RNA sequences (Papanicolaou *et al.,* 1984). Clearly, using this or a similar approach, any number of low-energy solutions can be generated. The major drawback is speed: Sequences longer than a few hundred bases take too long to analyze.

A stem-and-loop method based on the supposition that stems tend to form in order of descending equilibrium constants does not suffer from this size limit, however, since it considers only a minute subset of all conceivable structures (although a different subset than the global energy minimum method does). Thus, Martinez (1984) has developed a program that, at every point in the folding simulation, calculates an equilibrium constant for the formation of each as of yet unfolded stem compatible with the partial structure already formed; these equilibrium constants are estimated both from the stacking energies of the stem in question, and from the sizes of the loops on either side of it. This makes the folding process cooperative in the sense that two stem halves that are far apart in the linear sequence (and thus are prevented initially from pairing by a large unfavorable loop entropy) may be brought close together by more favorable stems forming in the intervening segment. If the remaining stem having the largest equilibrium constant is always picked, a unique structure will result. However, by picking stems randomly with the probability of being picked proportional to the equilibrium constant, a family of probable structures can be generated. This algorithm is very fast, with execution times proportional to N^2, and can be used on sequences as long as 30 kb (Mount, 1984). Whether or not the structure(s) generated have any biological relevance of course depends heavily on the correctness of the particular underlying folding scheme—the paucity of reliable experimental data on the folding and ultimate structures of RNA molecules makes an evaluation difficult.

A neat extension of these ideas has been put forward by Mironov *et al.* (1985). Rather than focusing on the equilibrium constants of stem formation, these authors try to estimate the kinetic constants for the association and dissociation of all possible stems at any given point in the simulation. Since in this scheme stems cannot only fold but also unfold, any unproductive cul-de-sac in the folding tree can be exited, and different structures can interconvert into one another. During the course of a Monte Carlo simulation, where the probability of a stem being picked is proportional to its association or dissociation rate, quickly formed local structures can refold into more stable but less quickly formed ones, and at longer times a stable equilibrium distribution will result. The method gives good results when applied to tRNAs (80 bases long), but presumably very quickly saturates the computer when applied to longer sequences.

Another way to use a list of candidate stems is to build the structure interactively, with the computer calculating stem-and-loop stabilities and displaying partial structures, but with all decisions concerning which stems to include left to the user. Available information on single- and double-stranded regions identified experimentally can easily be incorporated at the outset (Auron *et al.,* 1982; Stüber, 1986).

Finally, when many homologous sequences are known, an obvious way to proceed is to try to find common consensus folds. Trifonov and Bolshoi (1983) thus constructed dot matrices for a large number of 5S RNAs and superimposed them to find dyad symmetries common to all sequences in the group. They were able to find conserved inverted repeats that allowed the construction of two different folds—one of which has a tertiary interaction between two loops—and suggested that these may interconvert *in vivo.*

D. tRNA Genes

All tRNAs have a well-defined structure—the cloverleaf—and particular bases almost always appear at specific positions in this structure.

Staden (1980) has developed a search procedure that first looks for acceptable cloverleaf folds, and then tests all candidates against a supplied list of conserved bases. The distances between the stems are fixed at the outset; the minimal base-pairing in the individual stems and the allowed loop sizes are all entered by the user. An intron in the anticodon loop can be allowed, and its position will be indicated on the final display.

A similar program has been described by Paolella and Russo (1985); they also include a routine that searches for matches to the RNA polymerase III split promoter consensus RRYNNARYGG$(N)_m$ GTTCRANNC (R = purine, Y = pyrimidine, N = any base, and $m = 33-43$).

In conclusion, RNA folding is a favorite with molecular biology hackers: Base-pairing rules are conceptually simple; dynamic programming ideas from the sequence alignment field can be readily applied; and RNA structures are potentially important in many situations. In a sense, the results obtained so far are a little disappointing from the biological point of view, since one apparently cannot trust that the structures generated by the computer are anything like the ones found *in vivo*—not even for a simple and well-studied molecule such as tRNA. On the other hand, to the extent that one has additional information on the secondary structure of a particular RNA, the methods described above will greatly facilitate the task of finding those low-energy structures that are compatible with the data.

VIII. DNA GEOMETRY

In discussing DNA function and protein–DNA interactions, we have so far said nothing about the geometry of the DNA chain, i.e., its local and global structural parameters. Local variations in the precise positioning of the base pairs may influence the fit between amino acid side-chains and their binding sites on the DNA; the overall bending of a DNA segment may be important for the formation of nucleosomes; and gross structural alterations such as B- to Z-DNA transitions obviously could affect many functions of the double helix.

Theory and experiment in this area have just begun to shed light on these properties of DNA. Some simple rules for estimating the degree of local structural deviation from the ideal B-DNA geometry of any given nucleotide sequence are available, and have been used to characterize various DNA signals in terms of local structure rather than base sequence. Consistent patterns of structural deviations have been postulated to produce bends in the double helix, and can be localized with the aid of a computer. Algorithms for scanning sequences for likely Z-DNA tracts are also available.

A. Base Sequence and Helix Structure Variation

When Dickerson and his colleagues solved the X-ray structure of the DNA dodecamer CGCGAATTCGCG in the B form, they observed several sequence-dependent departures from the ideal regular Watson–Crick helix (Drew *et al.*, 1981). Calladine (1982) showed that these irregularities could be explained as the result of steric repulsion between purine bases (A,G) in consecutive base pairs but on opposite strands, and Dickerson (1983) later developed this idea into a simple quantitative scheme for calculating the variation in local geometry directly from the base sequence (or rather from the purine–pyrimidine pattern).

Calladine's model starts from the observation that the base pairs in B-DNA are not perfectly flat; instead they are tilted in relation to each other as in a propeller—they have a propeller twist (Fig. 4.20). In this way, each base increases the stacking interactions with its neighbors in the same strand, but at a price: Neighboring purines on opposite strands will clash either in the minor or major groove depending on whether the 5′→3′ sequence is pyrimidine–purine (Y-R) or purine–pyrimidine (R-Y) (Fig. 4.20a,b).

These steric repulsions can be relaxed in four ways:

1. By rotating one of the base pairs around the helical axis, decreasing the twist angle t_g (Fig. 4.21a).
2. By rolling the twisted base pairs as rigid units around their long axes away from the side where the clash is, increasing or decreasing the roll angle.
3. By sliding the base pairs along their long axes, decreasing the R-R overlap and changing the torsion angles in the phosphate backbone, (Fig.4.21b).
4. By decreasing the propeller twist, flattening the base pairs.

It turns out that all of these maneuvers are deployed, to a greater or lesser degree, depending on the base sequence. Also, clashes in the minor groove are more severe than major groove clashes for a given

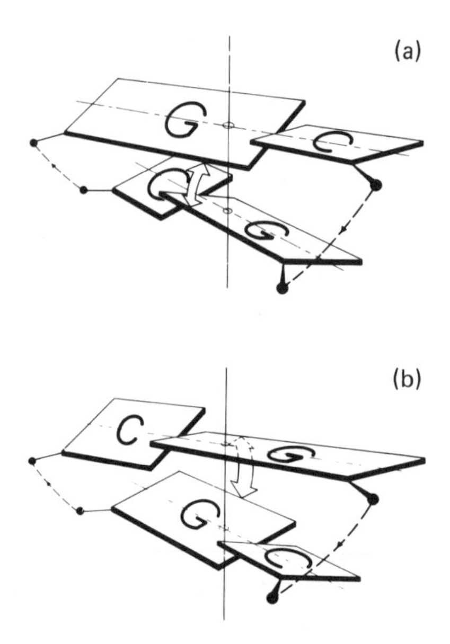

Fig. 4.20. Schematic drawing of how pyrimidine–purine (a) and purine–pyrimidine (b) nearest neighbors in B-DNA induce, as a consequence of the propeller twist, a purine–purine clash in the minor and major grooves, respectively. (From Dickerson, 1983.)

propeller twist. From an analysis of the Dickerson dodecamer data, it appeared that a simple set of rules could be used to quantitate the perturbations with a high degree of confidence (Dickerson, 1983). Thus, four sum functions $\Sigma_1 - \Sigma_4$ corresponding to the variation in twist, roll, torsion, and propeller twist angles can be calculated for each trimer or tetramer as follows:

1. Twist and roll angles
   a. x-R-Y-x: major groove clash. The R-Y twist angle is rotated by an amount -2; the x-R and Y-x twist angles are rotated by $+1$ (one unit corresponds to 2.1°). The net helical twist thus remains constant. The roll-angle rotations can be similarly defined as $+1, -2, +1$, with one unit corresponding to a 1.1° rotation.
   b. x-Y-R-x: minor groove clash. This clash appears from the experimental data to be twice as severe; the twist angle rotations imposed are $+2, -4, +2$, and the roll angle rotations are $-2, +4, -2$.

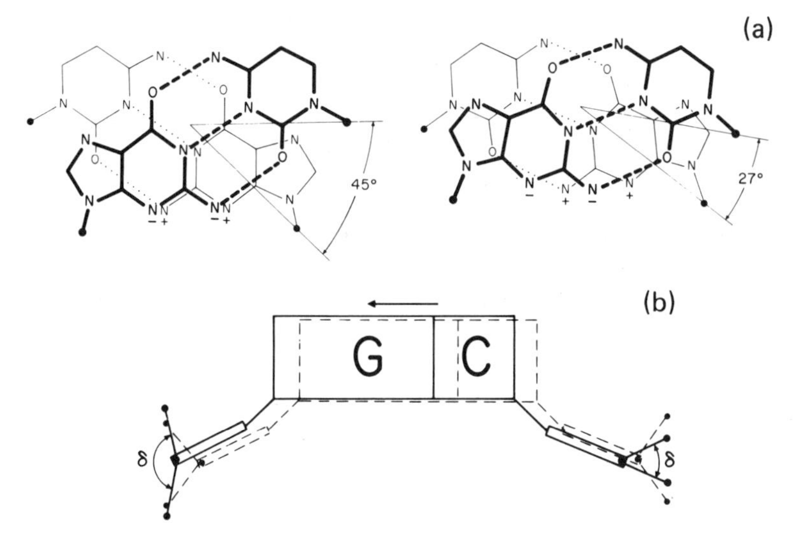

Fig. 4.21. (a) A decrease in the twist angle t_g can relieve a minor groove clash, cf., Fig. 4.20a. (b) In-plane sliding and concomitant change in the backbone torsion angle (δ) can also lessen the purine–purine overlap. (From Dickerson, 1983.)

2. Torsion angle and propeller twist
   a. R-Y: Torsion angle change is $+1$, -1; one unit is 15.6°. Propeller twist change is -1, -1; one unit is 3.6°.
   b. Y-R: Torsion angle change is -2, $+2$. Propeller twist change is -2, -2.

From these rules, the resulting perturbations in the four angles are calculated as shown in Fig. 4.22. The correlations between the calculated and observed variations are high, except for the propeller twist. Surprisingly, much more refined all-atom energy calculations give essentially similar results as these rules of thumb (Tung and Harvey, 1986).

Nussinov and co-workers have developed a whole sequence analysis package based on the Calladine–Dickerson rules (Shapiro *et al.*, 1986), and have analyzed a number of functionally important DNA sequences from this perspective. Enhancer elements (Nussinov *et al.*, 1984a), DNAase I hypersensitive sites (Nussinov *et al.*, 1984b), heat shock gene promoters (Nussinov and Lennon, 1984), and oligomer frequencies in eukaryotic DNA (Lennon and Nussinov, 1985) are

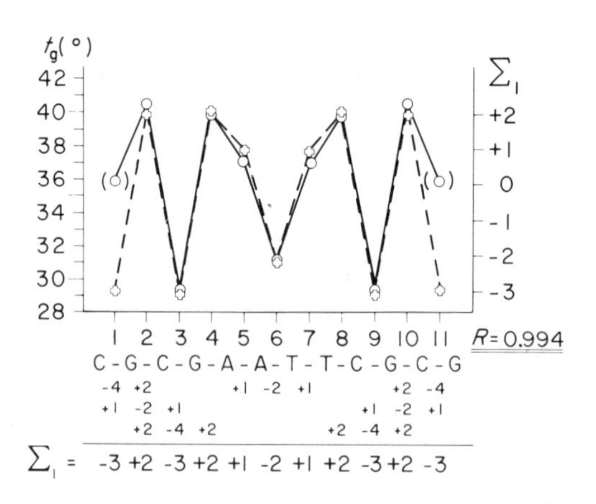

Fig. 4.22. Comparison of the observed twist angles, t_g, in the Dickerson dodecamer and the expected variation calculated from the Σ_1 sum function (as illustrated below the chart). (From Dickerson, 1983.)

some of the instances where variations in local geometry have been claimed to be important.

B. DNA Bending

If short stretches of DNA of specific sequence are repeated with a period close to the 10.5-bp pitch of the double helix, any local bend in the helical axis will add up to produce a persistent curving of the whole molecule. With this in mind, Trifonov and Sussman (1980) analyzed SV40 DNA, looking for any periodicities in the distribution of dinucleotides. Besides a 3-base repeat (and multiples thereof), they found a tendency for certain dinucleotides, AA (or the complementary TT) in particular, to be repeated with a period close to 10.5 bp. This suggested that sections of DNA may be bent or curved as a result of a wedge angle inherent to AA dinucleotides, possibly to make it easier to form the bent and supercoiled DNA in the nucleosomes.

Later experimental studies have confirmed the importance of the AA dinucleotide [in the form of short poly(A) tracts] and the 10.5-bp repeat distance; it appears, however, that the putative AA wedge cannot explain the results. Instead, it now seems likely that a short poly(A) tract, typically 5–6 residues long, has a geometry that is not bent but that nevertheless differs slightly from the normal B form; the bend is formed mainly at the 3′ junction between the poly(A) tract and the adjoining sequence (Ulanovsky *et al.,* 1986; Koo *et al.,* 1986; Hagerman, 1986). Bending has also been observed upon binding of the so-called heat shock transcription factor to its target site on DNA (Shuey and Parker, 1986), implying that local sequence patterns either opposing or facilitating bending might modulate the efficiency of protein–DNA interactions.

The concept of bend-promoting sequences with the appropriate periodicity has been used to devise a method for finding likely nucleosome binding sites (Mengeritsky and Trifonov, 1983). The tendency of the 10 different dinucleotides to be repeated with a period of 10.5 bp is quantified in terms of a matrix of bendability: The frequency of each kind of dinucleotide in each position in an 11-residues-long "archetypal" bend-promoting helical turn is taken as a weight matrix, and the sequence to be analyzed is scanned with this matrix as described previously (Section II,A,2). Since about 145 bases fold into the nucleosome core particle, successive 145-residue sections of the resulting, highly irregular, curve are then filtered into a periodical component of period 10.5 bp, and an irregular, nonperiodical component. The strength or amplitude of the periodical component for each 145-residue stretch is taken as a measure of the probability of finding a nucleosome at that position. The authors report that the method picks out most of the known nucleosome sites in a rather small number of examples, but it is not clear just how reliable the results are.

C. Z-DNA

Z-DNA, i.e., a left-handed double helix with a characteristic zigzag phosphate backbone of alternating *syn* and *anti* base conformations, has been detected experimentally both in negatively supercoiled circular DNA molecules and in chromosomal DNA. Its biological significance remains conjectural, but it has been proposed to be involved in gene regulation.

The B- to Z-DNA transition takes place most easily in segments of high GC content with long perfectly alternating purine–pyrimidine (R-Y) tracts, but the detailed requirements and the exact dependence on the degree of supercoiling are not known with precision (McLean *et al.,* 1986).

Nevertheless, a simple search for alternating R-Y tracts can possibly give some useful information related to the probability of B- to Z-transitions under appropriate conditions. Staden (1984b) includes such a routine in his package, and Vass and Wilson (1984) have written a program called ZSTATS that calculates the number of alternating R-Y runs in a window of 20 residues that is moved along the sequence, and estimates the probability that the observed number of runs could be produced by chance alone. Neither give any figures on how well these programs can identify known Z-DNA tracts.

A slightly more advanced program by Konopka *et al.* (1985) takes the nucleotide content as well as R-Y alternation into account. This program first finds all perfectly alternating fragments of length 4 bp or more, then constructs all quasi-alternating fragments of length 7 bp or more by concatenating all neighboring perfectly alternating tracts no more than 1 bp apart, and finally accepts only those quasi-alternating fragments that have less than 30% alternating A-T runs. The method finds almost all of the antibody-mapped Z-DNA sites in plasmid pBR322, phages ϕX174 and PM2, and in the SV40 genome; it predicts only a small number of additional sites that have not been detected experimentally.

The theoretically most well-founded algorithm now available is the Z-hunt program of Ho *et al.* (1986). Here, the B- to Z-DNA transition is modeled by the appropriate partition function from statistical physics and estimated B- to Z-DNA transition energies for all possible nearest neighbor pairs. For stretches of 16–24 bp, the degree of negative supercoiling necessary to induce the formation of 1 bp of Z-DNA in the stretch when placed in a circular plasmid the size of pBR322 is calculated. This is then converted into a Z score, which is defined as the number of bases in random DNA that must be searched on average to find a sequence that is as good or better at forming Z-DNA than the segment analyzed. The strength of this method is that it makes no *ad hoc* assumptions about which patterns of bases can and cannot form Z-DNA; the rules are embodied in the nearest neighbor transition energies and can be determined experimentally. The success rate is hard to estimate, in no small part because of the difficulties involved in mapping Z-DNA tracts in natural DNA.

D. Melting of DNA and Stability Domains

If the energetics of the B- to Z-DNA transition is only partially understood, the folding–unfolding process of B-DNA is much more well characterized, and will be the last kind of sequence-dependent process treated in this chapter.

The helix-coil transition induced by, e.g., thermal melting is highly cooperative, and can be described accurately by the formalism of statistical mechanics; in particular by the so-called one-dimensional Ising model (Wada *et al.,* 1980; Wartell and Benight, 1985; Wada and Suyama, 1986). Basically, neighboring base pairs are stabilized by stacking interactions, and are destabilized by electrostatic interactions between the backbone phosphate groups as well as by the entropy decrease resulting from the freezing of internal motions in the stacked base pairs. Two neighbors can either stack or not, i.e., melting of individual base pairs can be described as a two-state process. Using various approximations, the equations describing the cooperative melting of long DNA chains can be solved in a number of ways, but the most commonly used algorithm for calculating melting curves is the Poland–Fixman–Freire scheme (Fixman and Freire, 1977). To understand this method, a good amount of statistical mechanics is re-

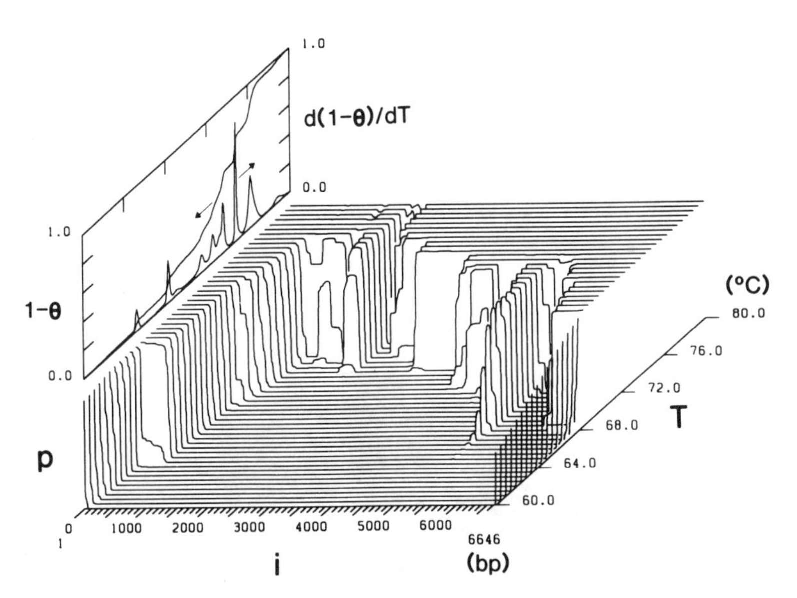

Fig. 4.23. Calculated melting surface of a 6646-bp DNA. *T* is the temperature, *i* is the position in the sequence, *p* is the base-pairing probability (zero at top, one at bottom). The melting profile (the helix content versus *T,* top left-hand side), as it would look in a temperature denaturation experiment, is obtained by projecting the melting surface onto the *T-p* plane. (From Wada and Suyama, 1986. Reprinted with permission from *Progress in Biophysics and Molecular Biology,* Vol. 47, Local stability of DNA and RNA secondary structure and its relation to biological functions. Copyright 1986, Pergamon Journals, Ltd.)

quired, and I will simply summarize the present state of the art by noting that melting surfaces, such as shown in Fig. 4.23, can be calculated for sequences of up to $\sim 10^4$ bp. The agreement between the theoretical calculations and experimentally observed melting curves is generally good but not perfect (Wartell and Benight, 1985).

From the biological point of view, the most interesting aspect of these methods, besides their obvious application to DNA–DNA and DNA–RNA hybridization experiments [where calculation by hand often is sufficient (Breslauer *et al.,* 1986)], is the possibility that patterns of local stability—DNA "breathing"—may be important for gene function. Thus, Wada and colleagues have suggested that genes or exons may be selected such that the double helix is of uniform stability throughout their lengths, with regions of sharp changes in the stability corresponding to gene boundaries (Wada and Suyama, 1986).

Chapter 5

Protein Sequences: What You Can Do With Your Sequence Once You Have It

It is clearly beyond the scope of this book to review the field of protein structure and folding; there are excellent books dedicated wholly to this task (Schulz and Schirmer, 1979; Creighton, 1983). Nevertheless, since most of the theoretical prediction methods to be described below focus on structural aspects of proteins, I will give a very brief background on how elements of secondary structure are thought to form and pack together. I will also provide a short introduction to the field of membrane proteins.

I. RULES OF PROTEIN STRUCTURE

The basic building blocks of proteins, the α-helix and the β-sheet, are of course inextricably linked with the name Linus Pauling. I will quote his own recollection of how he found the α-helix at length (L. Pauling, personal communication), because it provides an antidote to the belief (in part no doubt fostered also by this book) that all significant discoveries require sophisticated technology; sometimes sophisticated ideas alone can do the trick.

> In the spring of 1948, during the period when I was Eastman Professor in Oxford University, England, I spent a few days in bed with a bad cold. After a couple of days of reading I became bored, and decided to try again to determine the structure

> of the polypeptide chains in α keratin. I took a sheet of paper and drew a projection of a polypeptide chain on it, making the bond lengths and bond angles approximately correctly, except that the bond angle of the α carbon atom was made 180°. I then folded the sheet of paper along a line through each α carbon, connecting the peptide groups, to such a solid angle (dihedral angle) as to make the angle between the bonds equal to 109.5°, the tetrahedral angle. I repeated the folding along parallel lines, on the assumption that all of the amino-acid residues are equivalent. I looked to see if hydrogen bonds could be formed between the carbonyl group of one residue and the NH group of a residue farther along the chain. When the interatomic distances were not satisfactory for this hydrogen bond, I changed the direction of the line of folding, until the right angle was found. It turned out that the helical structure that resulted had 3.6 residues per turn. It is the structure of the α-helix, which is found in the α keratin and in a great many other proteins, including most of the globular proteins.

After Pauling, and after it became possible to solve the three-dimensional structure of proteins by X-ray diffraction, protein structure and folding have continually been in the forefront of molecular biology. A number of regularities and rules describing the folded structure have come to light, although we are still a far cry from being able to predict the full three-dimensional structure of a molecule from its amino acid sequence alone. According to Chothia (1984), the structure of soluble, globular proteins can be understood in general terms as follows: The principle underlying the structure of helices, sheets, and turns is the simultaneous formation of hydrogen bonds by buried peptide groups and the retention of single residue conformations close to those of minimum energy. The shape of the helix and sheet surfaces make these structural elements pack together in a small number of relative orientations. The links between secondary structures tend to be right-handed and short, and do not form knots. As a consequence, proteins usually fold to give secondary structures arranged in one or a few common patterns or folding units: $\alpha\alpha$, two antiparallel packed helices; $\beta\beta$, two antiparallel sheet strands; and $\beta\alpha\beta$, a helix packed against two adjacent parallel sheet strands. Roughly, globular proteins can be divided into four classes: all-α; all-β; α/β, built from $\beta\alpha\beta$ units; and $\alpha + \beta$, with helix and sheet regions segregated. The stability of protein structures arises mainly from a delicate balance between the reduction in the apolar surface area accessible to solvent that occurs on folding (and the simultaneous formation of hydrogen bonds), and the entropy loss associated with the formation of a well-defined structure.

The propensity of a given stretch of chain to fold into either a helix, a β-strand, or a turn is thus primarily dependent on (i) the preferred conformations of its constituent residues, and (ii) the packing quality of the surface formed. The relative success of secondary structure prediction schemes that take only local and semilocal sequence patterns into account is a consequence of this local character of the folding forces. Beyond these generalities, however, the detailed physics of protein folding is still rather vaguely defined (Gō, 1983b).

The structure of membrane proteins, on the other hand, is determined to a large extent by the apolar nature of the interior of the bilayer membrane. Basically, a protein can solve the "compatibility problem" in one of two ways: It can have long, apolar α-helices traversing the membrane, or it can form a closed β-structure (a barrel) with an apolar exterior facing the membrane and a central channel lined with more polar residues (see review by Eisenberg, 1984). Bacteriorhodopsin, and the reaction center complex from *Rhodopseudomonas viridis* [the structure of which was recently solved by X-ray diffraction (Deisenhofer *et al.,* 1985)] are examples of the first kind; porin, an outer membrane protein from *Escherichia coli,* exemplifies the second (Vogel and Jähnig, 1986).

II. SECONDARY STRUCTURE PREDICTION

Historically, the study of the physical chemistry of synthetic polypeptides provided the first indication that amino acids differ in their intrinsic abilities to form secondary structures (Schulz and Schirmer, 1979). When the first X-ray structures of globular proteins became available, it was soon realized that individual residues in real enzymes tend to have the same conformation as they have in their homopolymeric form; the correlation is far from perfect, however.

In the early 1970s a number of workers derived statistical propensities such as "α-forming" and "β-forming" abilities from the X-ray structures. These scales were then elaborated into full-blown empirical prediction schemes such as the well-known methods of Chou – Fasman and Garnier – Robson. A second approach, based on principles of residue packing rather than statistics, is exemplified by Lim's *a priori* method.

By now, the number of known X-ray structures has grown sufficiently to make it possible to find reasonable homologies between virtually any short oligopeptide and corresponding segments of known structure; this can be exploited by using the secondary structures of the segments in the X-ray data bank to predict the structure of the probe peptide. Another recent line of research tries to formulate folding rules in terms of specific patterns of amino acids and segments of secondary structure using the tools of artificial intelligence.

Here, I will describe each of these methods in some detail. As a somewhat discouraging end to the story, however, the final section on the reliability of secondary structure prediction schemes will show that the overall success rates of the different methods is not what one would call impressive.

A. Which Residues Prefer What Conformation?

Starting from known X-ray structures and a consistent definition of the helical, sheet, turn, and random coil states, one can simply count the number of times a given kind of residue i appears in a given conformational state j, n_{ij}. If N_j is the total number of residues in state j, N_T is the total number of residues in the whole sample, and n_i is the total number of residues of type i in the sample, one can define the statistical tendency [called the conformational parameter by Chou and Fasman (1974a)] for residue type i to appear in conformation j as

$$P_{ij} = (n_{ij}/n_i)/(N_j/N_T) \tag{5.1}$$

i.e., as the proportion of i residues that appear in state j normalized by the proportion of all residues that appear in state j.

P_{ij} values as compiled by Chou and Fasman (1978a) and Levitt (1978) are shown in Table 5.1. From these tabulations, residues can be classified according to their secondary structure propensities (Table 5.2). Note that although the Chou–Fasman and Levitt data agree in a general sense, some residues are classified differently in the two schemes.

Beyond simple overall frequencies, it turns out that some residues show marked preferences for either the amino- or carboxy-terminal ends of segments of secondary structure. This is particularly so for the helical conformation: Pro is rarely found except among the first three amino-terminal residues; Asp and Glu also prefer the amino-terminus to the carboxy-terminus; and the converse is true for Arg and Lys (e.g., Chou and Fasman, 1974a; Argos and Palau, 1982).

TABLE 5.1. **Secondary Structure Propensities**

Helix			Sheet			Turn		
Residue	C&F[a]	L[b]	Residue	C&F[a]	L[b]	Residue	C&F[a]	L[b]
Glu	1.51	1.44	Val	1.70	1.49	Asn	1.56	1.28
Met	1.45	1.47	Ile	1.60	1.45	Gly	1.56	1.64
Ala	1.42	1.29	Tyr	1.47	1.25	Pro	1.52	1.91
Leu	1.21	1.30	Phe	1.38	1.32	Asp	1.46	1.41
Lys	1.16	1.23	Trp	1.37	1.14	Ser	1.43	1.32
Phe	1.13	1.07	Leu	1.30	1.02	Cys	1.19	0.81
Gln	1.11	1.27	Cys	1.19	0.74	Tyr	1.14	1.05
Trp	1.08	0.99	Thr	1.19	1.21	Lys	1.01	0.96
Ile	1.08	0.97	Gln	1.10	0.80	Gln	0.98	0.98
Val	1.06	0.91	Met	1.05	0.97	Thr	0.96	1.04
Asp	1.01	1.04	Arg	0.93	0.99	Trp	0.96	0.76
His	1.00	1.22	Asn	0.89	0.76	Arg	0.95	0.88
Arg	0.98	0.96	His	0.87	1.08	His	0.95	0.68
Thr	0.83	0.82	Ala	0.83	0.90	Glu	0.74	0.99
Ser	0.77	0.82	Ser	0.75	0.95	Ala	0.66	0.77
Cys	0.70	1.11	Gly	0.75	0.92	Met	0.60	0.41
Tyr	0.69	0.72	Lys	0.74	0.77	Phe	0.60	0.59
Asn	0.67	0.90	Pro	0.55	0.64	Leu	0.59	0.58
Pro	0.57	0.52	Asp	0.54	0.72	Val	0.50	0.47
Gly	0.57	0.56	Glu	0.37	0.75	Ile	0.47	0.51

[a]Chou and Fasman (1978a).
[b]Levitt (1978).

TABLE 5.2. **Conformational Preferences of Amino Acids**

Classification		Class	Amino acids
Chou and Fasman (1978a)			
Strong helix formers		(Ha):	Glu, Met, Ala, Leu
Weak formers		(ha):	Lys, Phe, Gln, Trp, Ile, Val
Indifferent formers		(Ia):	Asp, His
Indifferent formers		(ia):	Arg, Thr, Ser, Cys
Weak helix breakers		(ba):	Tyr, Asn
Strong breakers		(Ba):	Pro, Gly
Strong sheet formers		(Hb):	Val, Ile, Tyr
Weak formers		(hb):	Phe, Trp, Leu, Cys, Thr, Gln, Met
Indifferent formers		(ib):	Arg, Asn, His
Weak sheet formers		(bb):	Ser, Gly, Gly
Strong breakers		(Bb):	Pro, Asp, Glu
Levitt (1978)			
Helix	favoring	(ha):	Ala, Leu, Met, His, Glu, Gln, Lys, (Cys)
	indifferent	(ia):	Val, Ile, Phe, Trp, Asp, Asn, Arg
	breaking	(ba):	Tyr, Thr, Gly, Ser, Pro
Sheet	favoring	(hb):	Val, Ile, Phe, Tyr, Thr, (Trp)
	indifferent	(ib):	Ala, Leu, Met, His, Gly, Ser, Arg
	breaking	(bb):	Glu, Gln, Lys, Asp, Asn, Pro, Cys
Turn	favoring	(ht):	Gly, Ser, Asp, Asn, Pro
	indifferent	(it):	Gly, Gln, Lys, Tyr, Thr, (Arg)
	breaking	(bt):	Ala, Leu, Met, His, Val, Ile, Phe, (Trp, Cys, Arg)

B. The Chou–Fasman Method

In its original form (Chou and Fasman, 1974b), the Chou–Fasman (CF) method was performed by hand by noting to which class each residue in the sequence belongs (Ha, hb, etc.), and locating clusters of helix or sheet formers with the aid of a few simple rules. A more quantitative prediction scheme was also presented, where the P_{ij} values are used according to the following rules (Chou and Fasman, 1978a,b):

1. A cluster of four helical residues (Ha or ha) out of six residues along the sequence will nucleate a helix, with Ia counting as 0.5 ha. The helical segment is extended in both directions until α-tetrapeptide breakers with $< P\alpha > \; < 1.00$ are reached. Pro can only occur among the first three amino-terminal residues. Pro, Asp, Glu and His, Lys, Arg are incorporated, respectively, at the amino- and carboxy-terminal helical ends. Any segment with $< P\alpha > \; \geq 1.05$ and $< P\alpha > \; > \; < P\beta >$ is predicted as helical.
2. A cluster of three β-formers (Hb or hb) out of 5 residues along the

sequence will nucleate a sheet. The sheet is extended in both directions until β-tetrapeptide breakers with $< P\beta > \; < 1.00$ are reached. Any segment with $< P\beta > \; \geq 1.05$ and $< P\beta > \; > \; < P\alpha >$ is predicted as β-sheet.

3. When regions contain both α- and β-forming residues, the overlapping region is helical if $< P\alpha > \; > \; < P\beta >$, or sheet if $< P\beta > \; > \; < P\alpha >$. The helix and sheet boundary frequency tables (not shown here) are also used to decide between α- and β-structure.

Rules to predict turns were also given; these rely on amino acid frequency tables giving the frequencies f of all 20 residues in positions $i,\ldots,i+3$ in a 4-residue bend, as well as the P_t values from Table 5.1. At each position along the chain one calculates $< P_t >$ (average over 4 neighboring residues) and $p_t = f_i \times f_{i+1} \times f_{i+2} \times f_{i+3}$. A bend is predicted when $p_t > 7.5 \times 10^{-5}$ as well as $< P_t > \; > 1.00$ and $< P\alpha > \; < \; < P_t > \; > \; < P\beta >$.

As has been noted by many authors, these rules do not uniquely define an unambiguous algorithm, and various attempts to computerize the method have been made. Generally, in one or other of its automated forms, the CF method performs less well than in the hands of the authors themselves (Section II,G).

C. The Garnier–Robson Method

Garnier, Robson, and collaborators (Garnier *et al.*, 1978) developed an empirical method of secondary structure prediction that is similar to the CF method in that it exploits the statistical properties of known structures. The Garnier–Robson (GR) approach, being based on a consistent application of information theory, is "cleaner" from the theoretical point of view, and is unambiguous and easy to program. It has the added advantage that auxiliary information from, e.g., circular dichroism measurements can be used to bias the prediction.

The underlying idea is that the conformational state of a given residue is determined not only by the residue itself, but also by neighboring residues. In terms of information theory, this influence can be represented by directional information terms, $I(S_j;R_{j+m})$, that give the effect of the residue type R at position $j+m$ on the conformational state S ($=\alpha, \beta$, turn, or coil) of the residue at position j. Thus, if a strong helix former is found in position $j-1$, say, it is quite likely that residue j belongs to a helix (even if it is only a weak helix former itself); this translates into a large term $I(\alpha_j;R_{j-1})$. The term $I(S_j;R_j)$ represents the propensity for state S of residue j considered in isolation, analogous to the Chou–Fasman P value.

In practice, 8 residues on each side of residue j are considered to contribute significant information regarding its conformational state, and the likelihood that residue j adopts conformation S is calculated as

$$L(S_j) = \Sigma I(S_j;R_{j+m}) \tag{5.2}$$

where the sum is taken from $m = -8$ to $m = +8$. The predicted conformational state S^* of residue j is simply the one with the largest L value. Values for the $20 \times 17 \times 4 = 1360$ I values (20 amino acids, 17 positions, 4 conformational states) were obtained from a representative sample of proteins with known structure (Robson and Suzuki, 1976); for a large enough sample the I values are given by

$$I(S;R_m) = \ln[f(S,R_m)/(1 - f(S,R_m))] - \ln[f(S)/(1 - f(S))] \tag{5.3}$$

where $f(S,R_m)$ is the frequency with which residues of type R are found m residues distant from a residue in conformational state S (for $m = 0$, this is simply the frequency of residue type R in state S), and $f(S)$ is the total number of residues in state S in the whole sample.

One attractive feature of this formulation is that cooperative effects are included from the beginning; no averaging or *ad hoc* rules concerning the minimum length of a given kind of secondary structure or preferences for amino- or carboxy-terminal ends need to be invoked.

It is also possible to subtract decision constants, *DC*s, one for each type of secondary structure, from the L values. This will bias the prediction toward one or the other state, and can improve the prediction significantly when the approximate amounts of α- and β-structure are known from, e.g., circular dichroism or Raman spectroscopy. The *DC* values recommended by Garnier *et al.* (1978) are given in Table 5.3 (note that these constants are subtracted from the respective L values).

D. *A Priori* Predictions

As an alternative to the "statistical" approach, one might start from a general model of protein folding and construct prediction schemes based on first principles. In the relatively few cases that this has been tried, the periodicity of the regular secondary structures has been a point of departure. The periodic distribution of amino acids on an ideal α-helix is best appreciated from helical wheel or helical net diagrams (Fig. 5.1) where one is looking either along the helical axis or at a

TABLE 5.3. Decision Constants Used in the Garnier–Robson Prediction Scheme[a,b]

Percentage of secondary structure (helix or sheet)	DC_{helix}	DC_{sheet}
Less than 20%	158	50
Between 20% and 50%	−75	−87.5
Over 50%	−100	−87.5

[a]Garnier *et al.* (1978).
[b]Note that $DC_{turn} = DC_{coil} = 0$.

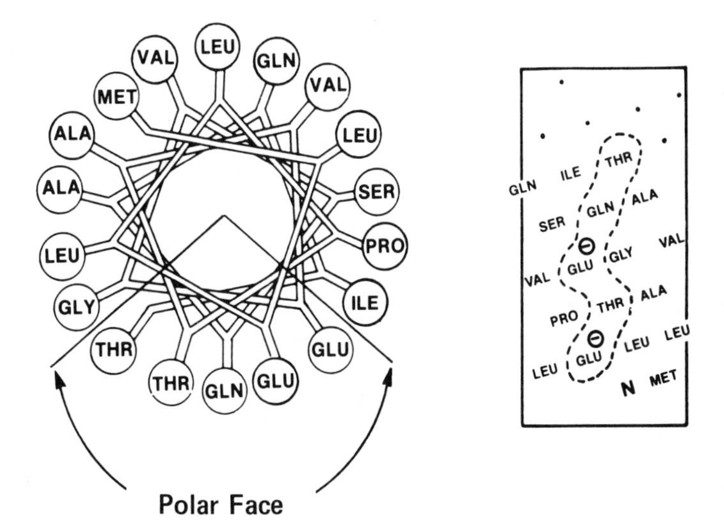

Fig. 5.1. Helical wheel and helical net representations of residues 1–18 from *Halobacterium halobium* bacteriorhodopsin. Note that the polar and charged residues tend to cluster on one face of the helix.

cylindrical projection that has been slit open and folded flat. The 3.6-residue pitch of the helix brings residues i, $i \pm 3$, and $i \pm 4$ together; similarly, in a β-sheet, residues i and $i \pm 2$ are neighbors on the same side of the sheet.

As briefly described in the introduction to this chapter, elements of secondary structure often pack together by virtue of complementary apolar surfaces. Thus, an α-helix should be possible to recognize from the regular appearance of apolar residues spaced 3 or 4 residues apart, whereas β-sheets will either be uniformly apolar—if they are completely buried—or have alternating polar and apolar residues. Indeed, helices and sheets in soluble proteins do display these characteristic patterns to a certain extent (Eisenberg *et al.*, 1984b).

The method of Lim (1974a,b) is an ambitious attempt to exploit regularities of this kind. He starts from a detailed examination of the way helices and sheets might pack together, formulates a model of protein folding, and derives a number of rather complex rules that make it possible to predict the secondary structure of a protein. It would carry us too far to attempt a detailed description of these rules; suffice it to say that potential helices are identified by requiring that they have patches of apolar residues in positions i, $i + 4$, or i, $i + 1$, $i + 4$, or i, $i + 3$, $i + 4$. Further rules and restrictions are then applied before the final prediction is arrived at.

Originally published as a pen-and-pencil method, Lim's scheme has later been computerized by Lenstra (1977).

E. Predictions Using Short Homologous Segments of Known Structure

With the rather large number of tertiary protein structures now available, it has become feasible to search these proteins for short stretches that are similar in sequence to segments of a protein of unknown secondary and tertiary structure. The hope is that oligopeptides of similar sequence will have similar secondary structures, and this seems to be borne out to some extent by the results so far.

In principle, two choices have to be made when one constructs an algorithm along these lines. First, a suitable measure of homology (similarity) must be defined and one must decide what degree of similarity should be required of the target sequences. Second, one must decide how to weigh the conformations of the segments found in the similarity search to arrive at the final prediction.

Three recent algorithms, those of Nishikawa and Ooi (1986), Levin *et al.* (1986), and Sweet (1986), although similar in outline, make somewhat different choices in these regards. Nishikawa and Ooi use a similarity score (see Chapter 6) based on six physical properties of the amino acids (polarity, turn propensities, mutability, partial specific volume, p*K* of α-amino group, and p*K* of α-carboxyl group) and calculate average scores over 11 neighboring residues, whereas Levin *et al.* and Sweet use different similarity matrices and calculate the match over 7 and 12 neighbors, respectively.

To predict the conformational state of residue *i* in the protein being analyzed, Nishikawa and Ooi and Levin *et al.* then count the number of times that the corresponding residues in the matching segments found in the database are in the α-, β-, and coil states. These counts are weighted by the similarity score calculated for each data base segment, and by the overall frequency of the respective state in the data base. Finally, residue *i* is predicted to be in the state that has the highest weighted count.

Sweet takes the 15 highest-scoring matches to each 12-mer and calculates the distribution of backbone dihedral angles for the residues corresponding to the one being predicted. This distribution is then weighted by multiplying it with an "average" distribution for the residue type in question calculated for the whole X-ray structure library, and the highest peak in the resulting distribution is taken as the prediction.

F. Pattern Recognition Methods

Another novel approach has been pioneered by Cohen and coworkers. Using artificial intelligence-type programming, they construct a hierarchical set of folding rules based on typical patterns of amino acids in a way reminiscent of Lim's method.

The way they identify turns gives a flavor of the idea (Cohen *et al.*, 1983). First, regions with a high density of hydrophilic residues (Asp, Glu, Gly, His, Lys, Asn, Pro, Gln, Arg, Ser, Thr) are identified by requiring at least 4 out of 5 or 5 out of 7 residues to belong to this group (or 3 out of 4, if the fourth is Tyr). These regions are labeled definite turns (DT). Then, weak turns (WT) are identified as follows: (i) if the length *L* between two DTs is less than 19 residues, no WT is assigned in this region; (ii) if $19 \leq L < 30$, one WT is possible among the central eight residues; a WT requires 3 out of 4, 4 out of 8, or 5 out of 9 hydrophilic residues; (iii) if $30 \leq L < 42$, WTs are possible among the ± 4 residues one-third and two-thirds along the way; (iv) if $42 \leq L$, four WTs are possible among the ± 4 residues one-, two-, and three-quarters along the way. This effectively partitions the sequence into segments of length 14 ± 4 residues. A slightly more involved procedure (Cohen *et al.*, 1986) is claimed to find more than 90% of the turns in a set of proteins of known structure.

Once the turns have been identified, the segments in between can be searched for helix or sheet candidates. If one knows that the protein is of a given structural class such as α/β, this knowledge can be used further to limit the prediction; a procedure that can identify likely α/β-regions has been described (Cohen *et al.*, 1983).

A drawback of the rule-based approach is that the method becomes rather difficult to grasp conceptually: The number of rules (and some rather complex, high-level rules at that) is simply too large to keep in mind. In a sense, there seems to be a trade-off between simple algorithms with few rules but many parameters (as in the traditional methods) and complex algorithms with many rules but few parameters. The latter methods seem to be more adaptable, however, since new rules and metarules can easily be added to the logical structure once they are recognized as being important. Its hierarchical nature is also well suited to recognize supersecondary structures such as the $\alpha\alpha$, $\beta\beta$, and $\beta\alpha\beta$ folding units.

G. Reliability of Secondary Structure Prediction Methods

A number of studies have tried to assess the performance of the secondary structure prediction methods. Different quality indices have been used by different authors (see Schulz and Schirmer, 1979), but the most popular is the simple percent correct score, Q, i.e., the percentage of residues that are placed in the correct structural class by the method in question. For a three-state prediction (α, β, and coil), a random assignment would give $Q = 33\%$; a random four-state prediction (α, β, turn, coil) would have $Q = 25\%$.

In their book, Schulz and Schirmer (1979) summarize the work in this area up to 1978 by stating that the conformational state of roughly two-thirds of all residues can be expected to be correctly predicted in a three-state prediction; no particular scheme seemed to perform better than the others.

More recently, however, Nishikawa (1983) checked the Chou–

Fasman, Garnier–Robson, and Lim methods on a sample of proteins whose tertiary structures had been solved *after* the publication of the methods. The results were disturbingly poor: For a three-state prediction, all three methods had Q scores around 55%, rather than the 68% quoted by Schulz and Schirmer. The four-state prediction yielded $Q = 45\%$. An attempt to improve the results by making a joint prediction using all three schemes simultaneously also failed. Similar results were also obtained by Kabsch and Sander (1983).

Nishikawa interprets this to mean that the original, more optimistic, results were peculiar to the data sets used in the construction of the methods, and he suggests that "what is now required may not be a partial improvement of the existing methods but a reconsideration of the basic methodology itself."

A further illustration of the shortcomings of the classical methods is provided by Garratt *et al.* (1985), who show that *internal* β-strands that are buried inside the folded protein are much more accurately predicted than more exposed strands (Q drops from 60% to 20% in going from the central to the peripheral strands in a three-state Garnier–Robson prediction). It thus seems clear that a more realistic and detailed understanding of the factors responsible for protein folding needs to be embodied in the secondary structure prediction methods before they will attain a sufficient quality to allow higher order structures to be predicted with any degree of confidence. Possibly, the homologous segments methods may perform slightly better than the old methods; Levin *et al.* (1986) claim a Q value slightly over 60% in a three-state prediction for a set of seven proteins not in the original data base.

III. SUPERSECONDARY STRUCTURE PREDICTION

Going beyond the secondary structure, supersecondary structures such as the $\alpha\alpha$, $\beta\beta$, and $\beta\alpha\beta$ domains could conceivably be recognized in the primary sequences. One can even imagine approaching the final tertiary structure by "docking" elements of secondary or supersecondary structure on the computer and evaluating the resulting folds by some set of criteria. This is still more of a research program than concrete, reliable algorithms (Blundell *et al.*, 1987), but I will nevertheless describe a couple of attempts to come to grips with problems of this sort.

Sternberg and colleagues have conducted a number of studies on how helices and sheets might coalesce into stable domains (reviewed in Sternberg, 1983) and ultimately into the native fold. These were based on the idea that the very large number of structures possible with a given set of secondary structure elements could be efficiently pruned by appealing to known regularities such as helix–helix or helix–sheet packing rules and the preferred topologies or handedness of connecting links. Indeed, the native structure was often found among the top few hundred out of some $10^6 - 10^7$ possibilities. Still, the number of nonnative competitors is clearly too large, and the presently available secondary structure prediction methods are far from accurate enough to provide good input data to the docking algorithms.

In the short run, a more promising, though less general, approach seems to be the application of pattern recognition methods for finding local domains such as the $\beta\alpha\beta$-fold. The rule-based artificial intelligence schemes of Cohen *et al.* have already been referred to; in one case, such a scheme has been used to discriminate between α/β- and non-α/β proteins, and to predict the most probable locations of the $\beta\alpha\beta$-units (Cohen *et al.*, 1983).

Another $\beta\alpha\beta$-finding program has been developed by Taylor and Thornton (1983, 1984). They first constructed an "ideal" $\beta\alpha\beta$ unit from an analysis of 62 such units of known structure, a (5-residue strand) — (5-residue coil) — (12-residue helix) — (5-residue coil) — (5-residue strand) pattern. This was subsequently used to scan Garnier–Robson secondary structure predictions in a search for regions with a reasonable fit to the ideal, and a new secondary structure prediction was made which, in regions of reasonable fit, was biased toward the ideal structure. The authors were able to locate 70% of all $\beta\alpha\beta$-units in their sample, and could record an average 7.5% improvement in the secondary structure prediction.

An extension along the same lines has been presented by Taylor (1986a), who has constructed consensus templates for all individual strands in the conserved immunoglobulin β-sheet, and has shown that these templates can find their respective strands with high accuracy and also discriminate immunoglobulins from other, unrelated proteins.

A similar template or "fingerprint" specific for the ADP-binding $\beta\alpha\beta$-fold in nucleotide-binding proteins has been used by Wierenga *et al.* (1986) to locate probable elements of this kind in the NBRF Protein Sequence Data Bank. The template specifies which amino acids are allowed in each of 11 positions in a 28- to 32-residues-long segment; e.g., positions 6, 8, and 11 must be Gly, positions 15 and 18 can be Ala, Ile, Leu, Val, Met, or Cys, etc. All segments with at most one mismatch to the template found in the database seem to have the ADP-binding fold.

The template and fingerprint methods have recently been generalized by Gribskov *et al.* (1987), who start out with a set of related sequences aligned on the basis of their 3-D structures. A "profile" weight-matrix is then constructed that reflects the pattern of amino acids found at each position in the aligned set and also assigns a weight (or penalty) for insertions and deletions at each position. New sequences are aligned with the profile matrix using a dynamic programming algorithm (see Chapter 6) that optimizes the number of matches by making appropriately weighted insertions and deletions. From the alignment, the best match not only to the sequences in the profile but to their common tertiary folding pattern is found. One use of this method is in trying to establish distant relationships between proteins —relationships that involve overall tertiary structure rather than precise amino acid homology—or in trying to predict the folding class of a new protein.

In this context, we might note the attempt by Vonderviszt and Simon (1986) to use nearest and next-nearest neighbor amino acid frequencies to locate domain boundaries from the primary structure. The rationale for the method is that, since the observed pair frequencies in a large sample of protein sequences should mostly reflect intradomain constraints, domain boundaries might appear as minima in a plot of the pair-frequency index versus sequence position. In a couple of cases this indeed seems to be the case, but other minima not corresponding to domain boundaries are plentiful in the plots as well. It is clearly too early to assess the usefulness of this method.

IV. PREDICTING THE STRUCTURAL AND FUNCTIONAL CLASS OF A PROTEIN

Faced with a protein sequence of unknown structure and function, we might settle for less than a full secondary structure prediction, i.e., it may be sufficient to classify the sequence in terms of folding type (all-α, all-β, etc.), function (globin, chromosomal protein, enzyme), or location (intra- or extracellular protein, membrane protein). This can be accomplished in a number of ways: We can search for homologous proteins in a protein sequence data bank (Chapter 6); we can estimate the total secondary structure content from a secondary structure prediction algorithm; we can look for specific sorting signals in the amino acid sequence (Section VI).

A more general way to allocate proteins to a number of suitably defined groups is to use some sort of multidimensional cluster analysis. The idea is to characterize each sequence ("object") by a number of global "features" or "attributes," such as its amino acid composition (20 attributes), its length, the number of runs of polar or apolar residues longer than some threshold, its mean hydrophobic moment (Section V,C), or whatever else one might care to try.

These n attributes will define a n-dimensional space, with each attribute corresponding to one coordinate. Every sequence is thus represented by a point in this space, and, if the attributes have been properly chosen, the cloud of points representing a large sample of different proteins will tend to cluster into a small number of denser clouds. If this happens, and if the denser clouds correspond to some natural classification such as structural class or location in the cell, one can obviously classify an unknown protein by calculating the values of its attributes and assigning it to the closest dense cloud in the n-dimensional space.

Once the attributes to be analyzed have been chosen, there are a number of more or less equivalent ways to perform the classification; these statistical details need not concern us here.

A number of attempts to classify proteins into one of the five structural classes all-α, all-β, α/β, $\alpha + \beta$, and irregular are on record. The attributes used include the 20 amino acid frequencies (Nakashima *et al.*, 1986), the amino acid frequencies plus the number of runs of 4 or more polar and apolar residues and the mean hydrophobic moment (Sheridan *et al.*, 1985), and hydrophobic moment with 2 and 3.6 residues periodicity plus percentages of α- and β-content estimated by the Garnier–Robson method (Klein and DeLisi, 1986). Generally, some 70% of the proteins analyzed are classified correctly when all five structural groups are considered. Since the Garnier–Robson secondary structure prediction method can be biased on the basis of a structural class assignment (Section II,C), it might be possible to improve its accuracy slightly by first determining the structural class in this way, and then set the appropriate decision constants.

Nishikawa *et al.* (1983a,b) have found that the 20 amino acid frequencies can be used to determine whether a protein is an enzyme or a nonenzyme, and whether it is intra- or extracellular. The intra- or extracellular location can be predicted with ~80% confidence (two-state prediction with random expectation = 60%); if one makes a four-state prediction (intracellular/extracellular enzymes/nonenzymes), 66% of the proteins in the sample are correctly classified.

Finally, Klein *et al.* (1984, 1986) have tried to classify proteins into the functional groups globins, chromosomal proteins, contractile system and respiratory proteins, enzyme inhibitors and toxins, enzymes, and all other proteins. The attributes chosen were average hydrophobicity, net charge, sequence length, and hydrophobic moment (Klein *et al.,* 1984), or a large number of attributes including these as well as amino acid frequencies, maximum hydrophobicity and variance in hydrophobicity, charge periodicity, hydrophobic runs, number of α- and β-forming residues, etc. (Klein *et al.,* 1986). When classification into the six groups listed above was carried out, some 75% of all sequences were assigned correctly.

Clearly, cluster analysis is in many ways subject to similar shortcomings as secondary structure prediction: One measures properties that are only indirectly related to the actual three-dimensional structure; the results depend more or less heavily on the data base or "training set" used to derive the statistical rules, and, indeed, one is still far from achieving 100% accuracy (except for very well-defined groups such as globins, but these have strong evolutionary relationships and proteins belonging to such groups would probably be identified equally well by homology searches). Cluster analysis is nevertheless potentially valuable as a tool to get rapidly an idea of what an open reading frame of unknown significance might in fact represent, or as a guide to setting decision constants in secondary structure prediction.

V. THE POLAR–APOLAR NATURE OF SOLUBLE AND MEMBRANE-BOUND PROTEINS

It is a commonplace observation in undergraduate chemistry that "like dissolves like," i.e., that molecules tend to prefer neighbors of their own kind to strangers with dissimilar chemical characteristics. In aqueous solution, apolar molecules or parts of molecules tend to aggregate such that the total apolar surface area exposed to water is minimized. Likewise, the structure of a globular protein is to a large measure dictated by the requirement that its polar and apolar residues should be exposed, respectively, on its surface and buried in its interior as far as possible.

The intuitive concept of a "hydrophobic bond" between apolar molecules, or a "hydrophobic effect," that accounts for such diverse chemical phenomena as the stability of lipid bilayers and the folding of large proteins, lies behind much of our present understanding of the physical chemistry of the living cell; yet, hydrophobicity has proved very difficult to quantify unambiguously. Clearly, it must in the final analysis be related to quantum mechanical properties of both solvent and solute (e.g., van der Waals interactions, hydrogen bonds, electrostatic energies) as well as to entropic effects stemming mainly from the relatively well-ordered water layer that forms around any apolar surface—all difficult problems in their own right that are far from completely understood.

Fortunately, chemists and, perhaps even more, molecular biologists are pragmatic people. There are scores of hydrophobicity scales in the literature, scales that have either been optimized with some given application in mind, or that have been derived from various experimental approaches. A brief introduction to the most popular ones is next, followed by sections on hydrophobicity and hydrophobic moment analysis, antigenic site and membrane protein structure prediction.

A. Scales of Hydrophobicity

1. *Experimental Scales*

The obvious first choice for obtaining a simple experimental scale is to measure the solubility difference or partitioning of a series of model compounds between water and some apolar solvent. The first such study of the hydrophobicities of the naturally occurring amino acids (Nozaki and Tanford, 1971) used ethanol or dioxane as the apolar medium. By subtracting the value for glycine, these authors arrived at a hydrophobicity scale expressing the free energy of transfer for a given amino acid side chain from organic solvent to water (Table 5.4). As an example of an updated version of this approach, Fauchère and Pliska (1983) have measured the partitioning of all 20 *N*-acetylamino acid amides in octanol/water at pH 7. More recently, Parker *et al.* (1986) have used high-performance liquid chromatography (HPLC) retention times for 20 model peptides of the composition Ac-Gly-X-X-$(Leu)_3$-

TABLE 5.4. **Hydrophobicity Scales**

	Source												
Residue	Nozaki	Bull	Wolfenden	Manavalan	Janin	Argos	von Heijne	Kyte–Doolittle	Eisenberg	Fauchère	Parker	Sweet	GES
Ala	0.5	−0.2	1.9	13.0	0.3	1.6	−1.0	1.8	0.25	1.52	2.1	−0.40	−1.6
Cys	—	−0.5	−1.2	14.6	0.9	1.2	−1.5	2.5	0.04	1.70	1.4	0.17	−2.0
Asp	—	−0.2	−10.9	10.9	−0.6	0.1	7.4	−3.5	−0.72	2.60	10.0	−1.31	9.2
Glu	—	−0.3	−10.2	11.9	−0.7	0.2	5.9	−3.5	−0.62	2.47	7.8	−1.22	8.2
Phe	2.5	−2.3	−0.8	14.0	0.5	2.0	−3.4	2.8	0.61	0.04	−9.2	1.92	−3.7
Gly	0.0	0.0	2.4	12.4	0.3	0.6	0.0	−0.4	0.16	1.83	5.7	−0.67	−1.0
His	0.5	−0.1	−10.2	12.2	−0.1	0.3	3.4	−3.2	−0.40	1.70	2.1	−0.64	3.0
Ile	2.6	−2.3	2.2	15.7	0.7	1.7	−2.5	4.5	0.73	0.03	−8.0	1.25	−3.1
Lys	—	−0.4	−9.5	11.4	−1.8	0.2	4.2	−3.9	−1.10	2.82	5.7	−0.67	8.8
Leu	1.8	−2.5	2.3	14.9	0.5	2.9	−2.4	3.8	0.53	0.13	−9.2	1.22	−2.8
Met	1.3	−1.5	−1.5	14.4	0.4	3.0	−2.7	1.9	0.26	0.60	−4.2	1.02	−3.4
Asn	—	0.1	−9.7	11.4	−0.5	0.3	2.9	−3.5	−0.64	2.41	7.0	−0.92	4.8
Pro	—	−1.0	—	11.4	−0.3	0.8	3.3	−1.6	−0.07	1.34	2.1	−0.49	0.2
Gln	—	0.2	−9.4	11.8	−0.7	0.5	2.4	−3.5	−0.69	2.05	6.0	−0.91	4.1
Arg	—	−0.1	—	11.7	−1.4	0.5	11.3	−4.5	−1.80	2.84	4.2	−0.59	12.3
Ser	−0.3	−0.4	−5.1	11.2	−0.1	0.8	1.5	−0.8	−0.26	1.87	6.5	−0.55	−0.6
Thr	0.4	−0.5	−4.9	11.7	−0.2	0.9	0.9	−0.7	−0.18	1.57	5.2	−0.28	−1.2
Val	1.5	−1.6	2.0	15.7	0.6	1.1	−2.0	4.2	0.54	0.61	−3.7	0.91	−2.6
Trp	3.4	−2.0	−5.9	13.9	0.3	1.1	−2.0	−0.9	0.37	−0.42	−10.0	0.50	−1.9
Tyr	2.3	−2.2	−6.1	13.4	−0.4	0.7	1.1	−1.3	0.02	0.87	−1.9	1.67	0.7

$(Lys)_2$-amide to construct yet another hydrophobicity scale based on partitioning equilibria.

Hydrophobic residues not only partition into apolar solvents, they also form films at the air–water interface, thus removing much of their apolar surface from contact with water. The strength of this tendency can be quantified by measuring the variation in surface tension with amino acid concentration, which in turn can be converted into a free energy of transfer from solution to surface (Bull and Breeze, 1974). In a similar vein, compounds directly mimicking the amino acid side-chains (hydrogen for Gly, isobutane for Leu, etc.) have been used to estimate the free energy of transfer of amino acid side chains from the vapor phase to neutral aqueous solution (Wolfenden *et al.,* 1979).

Clearly, one difficulty with these experimentally derived scales is to know how well they represent the environments experienced by the amino acid *in vivo:* The interior of a globular protein or the hydrocarbon core of a membrane. To this end, attempts have been made to construct hydrophobicity scales starting from the other end, i.e., from the final protein structure itself.

2. *Statistical Scales*

In the study of protein structure, the term hydrophobicity is used to describe the tendency of a given kind of amino acid residue to be sequestered inside the folded molecule. As the number of globular proteins with known three-dimensional structure has increased, it has become possible to obtain statistical estimates of this tendency. Central to this approach is the notion of accessible surface area, i.e., the area of a given residue or side-chain in a given protein that is exposed to solvent (see, e.g., Rose *et al.,* 1985a, for a precise definition), and efficient algorithms for computing this value from the atomic coordinates of a protein have been available for a long time. An important early result was the observation by Chothia (1974) that the accessible surface areas calculated for the central residue of Ala-X-Ala tripeptides correlates closely with the hydrophobicity values measured by Nozaki and Tanford (1971): each $Å^2$ of accessible surface area was found to correspond to approximately 24 cal/mol of transfer free energy, and each hydrogen bond that had to be broken upon transfer of a polar group into a apolar environment apparently lessened the hydrophobicity of the residue by about 1 kcal/mol. Frömmelt (1984) and Eisenberg and McLachlan (1986) have shown further that this correlation can be decomposed into separate contributions from each kind of atom in the residue transferred.

Truly statistical partition coefficients for the transfer of residues between the exterior and interior of proteins was derived by Janin (1979). He defined as "buried" all residues with accessible surface areas less than 20 Å^2 in his sample of globular proteins, and calculated for each kind of residue a free energy of transfer as $\Delta G = RT\ln$(number of buried residues/number of accessible residues) (Table 5.4). Rose *et al.* (1985b) have similarly calculated the mean difference in accessible surface area (corresponding to a transfer free energy) between the unfolded and folded states for the 20 amino acids in a sample of 23 globular proteins.

Rather than focusing on the residue itself, Manavalan and Ponnuswamy (1978) and Miyazawa and Jernigan (1985) have considered the neighbors of a given kind of residue in globular proteins of known three-dimensional structure. Manavalan and Ponnuswamy thus define the surrounding hydrophobicity or bulk hydrophobic character of a residue as the mean hydrophobicity of all residues found within an 8 Å radius of its center (not counting the residue itself). Miyazawa and Jernigan estimate contact energies (including hydrophobic as well as electrostatic and hydrogen-bond contributions) between all possible residue–residue pairs by analyzing the statistics of pair contacts in the known structures. More complicated schemes have also been considered, e.g., to divide the protein into concentric layers and study the distribution of residues between these layers (Guy, 1985).

Sweet and Eisenberg (1983), in an attempt to construct a hydrophobicity scale that can be used in the study of the evolution of related proteins, have converted the Eisenberg consensus scale (next section) into optimal matching hydrophobicities that reflect both the hydrophobicity and the observed rate of substitution between different residues (see Chapter 6).

The tendency of amino acids to like or dislike an apolar environment is perhaps even more decisive for the structure of membrane proteins than for soluble ones. Indeed, as will become clear later on, it may prove easier to come to grips with the forces shaping a membrane-buried protein segment than with those determining the folding of a globular molecule. A couple of scales specifically intended for the study of lipid–protein interactions are those of von Heijne (1981) and Engelman and Steitz (1981)—basically extensions of the surface area/

free energy correlation noted above—Argos *et al.* (1982) and Kuhn and Leigh (1985). The two latter derive membrane propensity parameters by dividing the fraction of a given kind of amino acid in a sample of membrane-buried segments by its fraction in a sample of globular proteins.

Guy (1985) and by Rose *et al.* (1985a) have correlated various scales; Rose *et al.* discern a Nozaki–Tanford group and a Wolfenden group of scales with rather poor correlation between the two groups.

3. *Aggregate Scales*

Once a sufficient number of similar but not identical scales exist, people will start to present aggregate scales based on the idea that averaging will tend to minimize unknown systematic errors present in the individual scales (and possibly because it is easier to calculate averages than to construct conceptually novel scales). This may well be true, but it could equally well be the case that averaging dilutes out some distinctive feature that makes one particular scale suitable in a particular application. In any case, a couple of aggregate scales have become extremely popular and are used routinely; this has the positive effect of making comparisons between different authors easier, and should be encouraged so long as none of the primary scales have been clearly shown to be superior in a given context.

By far, the most widely used aggregate scale is that presented by Kyte and Doolittle (1982). For most of the 20 amino acids the hydropathy value assigned by these authors is simply the mean of normalized versions of the Wolfenden solvation energy scale and two statistical scales similar to Janin's (Chothia, 1976). However, for a number of residues "subjective adjustments" were introduced, often after "futile or heated discussions" between the authors—a rare confession in the world of "objective" science.

A second scale that is sometimes used because of its close links with a method for calculating the so-called hydrophobic moment profile (Section V,C) of a sequence is Eisenberg's consensus scale (Eisenberg, 1984), an amalgamation of the Wolfenden, Janin, von Heijne (see Table 5.4), and Chothia (1976) scales.

The reader may well feel overpowered and helpless in the presence of such a profusion of scales: Which particular one should I choose? He can derive some reassurance from the experience not only of Kyte and Doolittle (1982) but also of others that, insofar as one is analyzing the hydropathy of a globular or membrane-bound protein, "the number in the second place of the hydropathy values is of little consequence to the hydropathy profiles."

B. Hydrophobicity Plots

A hydrophobicity plot, i.e., a graph displaying the distribution of polar and apolar residues along a sequence, is most often used either to spot putative membrane-buried segments and exposed parts such as antigenic determinants, or to compare distantly related sequences to see if some characteristic pattern of hydrophobicity has been preserved (Chapter 6). Basically, one scans the sequence with a moving window of a given length, and computes, for each position of the window, the mean hydrophobicity of all residues in the window. The hydrophobicity (or hydropathy) profile thus depends on one's choices of scale and window length.

Some representative plots for a globular and a membrane-embedded protein are shown in Fig. 5.2. Considering the crudeness of the hydrophobicity scales, the choice of scale has little effect on the results. The window length, on the other hand, is more important, and one should choose a length that corresponds to the expected size of the structural element one is looking for: A window of 5–7 residues seems appropriate for finding surface-exposed regions that might be antigenic sites (Hopp and Woods, 1981), whereas a length of 19–21 residues will make long, hydrophobic membrane-spanning segments stand out more clearly. In this regard, it is a little unfortunate that the original Kyte–Doolittle plots (Kyte and Doolittle, 1982) were made with a 7-residue window (although they noted that a length of 19 residues gave a better discrimination of membrane-spanning parts) since many authors presumably not very familiar with the original paper now seem to use a "7-residue default" option under all circumstances.

C. The Hydrophobic Moment

Hydrophobicity analysis can be extended to search not only for unbroken stretches of hydrophobic or hydrophilic residues but also to detect periodicities in the distribution of such residues. Any periodic secondary structure (i.e., an α-helix or a β-sheet) can be characterized by the number of residues per turn m, or alternatively by the angle δ at which the side-chains of successive residues project from the backbone:

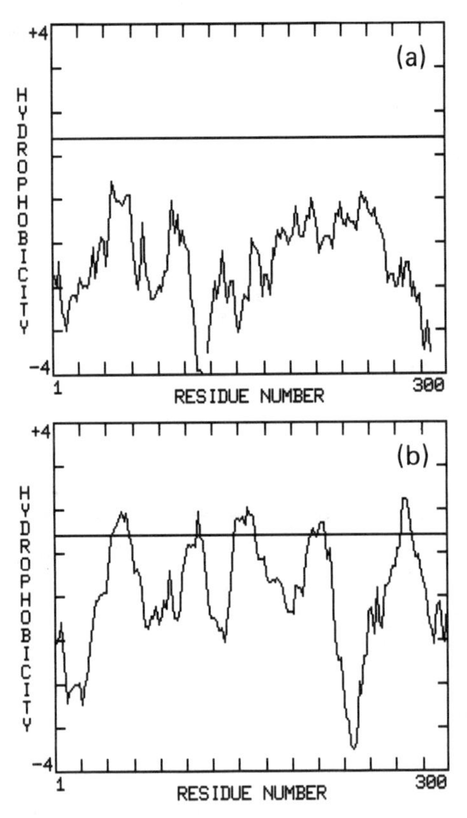

Fig. 5.2. Hydrophobicity plots of *E. coli* L-arabinose binding protein, a globular protein (a), and *R. viridis* reaction center M subunit, an integral membrane protein (b). The GES scale (Table 5.4) with a window size of 19 residues was used in both a and b. The horizontal line corresponds to a hydrophobicity value of +1.4.

$\delta = 2\pi/m$. Thus, for an α-helix $\delta = 100°$, a 3_{10} helix has $\delta = 120°$, and a β-structure is characterized by $\delta = 160-180°$.

For any value of δ, one can define a hydrophobic moment vector (Eisenberg *et al.*, 1984a):

$$\mu = \Sigma h_i \mathbf{s}_i \tag{5.4}$$

where h_i is the hydrophobicity of residue i and $\mathbf{s}_i$ is a unit vector pointing from the helical axis toward the α-carbon of residue i. The sum is taken over a given window (typically 11 or 18 residues long) that is moved successively along the sequence.

In most applications only the absolute value of μ is of interest; this is easily computed as

$$\mu = \{[\Sigma h_i \sin(\delta i)]^2 + [\Sigma h_i \cos(\delta i)]^2\}^{1/2} \tag{5.5}$$

A large value of μ for a given δ indicates that polar and apolar residues tend to segregate to opposite sides of the periodic structure in the segment under consideration. Thus, plots of μ versus the position of the window, or even contour plots of $\mu = f(\delta)$ versus sequence position (Finer-Moore and Stroud, 1984), can be used to detect elements of amphiphilic secondary structure.

It is also possible to analyze each helical face separately. Vogel and Jähnig (1986) calculate the mean hydrophobicity of one side of a 20-residue helix with residue i in the middle from the formula

$$H\alpha(i) = \tfrac{1}{10}[h_{i\pm 8} + h_{i\pm 7} + h_{i\pm 5} + h_{i\pm 4} + \tfrac{3}{4}h_{i\pm 3} + \tfrac{1}{2}h_{i\pm 1} + h_i] \tag{5.6}$$

and similarly for one side of a 10-residue β-strand with residue i in the middle:

$$H\beta(i) = \tfrac{1}{5}[h_{i\pm 4} + h_{i\pm 2} + h_i] \tag{5.7}$$

where h_i is the hydrophobicity of residue i.

In an amphiphilic helix or strand, $H\alpha$ or $H\beta$ will fluctuate wildly between neighboring positions, as one goes from the apolar to the polar and back to the apolar face. The advantage of this kind of analysis is that the hydrophobicity of the apolar face as well as the hydrophilicity of the polar face will be immediately apparent from the plot of $H\alpha$ or $H\beta$ versus sequence position; in a plot of μ, only the *difference* between the two sides is shown. On the other hand, the Vogel–Jähnig plots easily become rather cluttered, and are not amenable to smoothing procedures.

The hydrophobic moment has been generalized in another way by Schwyzer (1986), who attempts to calculate the orientation of an amphiphilic peptide relative to an interface separating apolar and polar media. He thus calculates a generalized amphiphilic moment **M**:

$$\mathbf{M} = \Sigma h_i \mathbf{R}_i \tag{5.8}$$

where $\mathbf{R}_i$ is the unit vector pointing from the *center of mass* of the helical peptide (not the helical axis) toward the α-carbon of residue i. The peptide is then predicted to be oriented such that **M** is perpendicular to the interface.

Finally, the tendency of a given peptide to form an amphiphilic helix in different environments can be estimated from helix–coil transition theory specifically incorporating terms that describe the stabilization/destabilization expected from pairs of charged residues on the polar face of the helix (Hamed *et al.,* 1983).

D. Predicting Antigenic Sites

Raising antibodies to defined proteins is an important step in many experimental investigations, as well as in the production of vaccines of various kinds. Often, however, the protein is available in insufficient quantities for immunization. If the sequence of the protein is known, one might be able to raise antibodies to synthetic peptides mimicking only a part of the molecule; the question is how to make an educated guess as to which peptide(s) would be the most likely to correspond to a true antigenic site on the intact protein.

First, it must be understood that most if not all antigenic sites on proteins are discontinuous, i.e., composed of residues that are far apart in the sequence, rather than continuous. The small number of sites that are continuous often correspond to loops and/or protruding regions on the protein's surface (Barlow *et al.,* 1986). These, then, are the kinds of elements one would like to be able to predict with confidence from the primary sequence—the identification of likely antigenic sites, i.e., protruding or highly mobile regions on the surface, from an X-ray structure is, not surprisingly, a lot easier (Thorton *et al.,* 1986; Fanning *et al.,* 1986; Novotny *et al.,* 1986).

Regions likely to be exposed on the surface of a folded protein obviously should be rich in polar and charged residues. Such hydrophilic clusters can be identified easily from hydrophobicity plots, and the methods that have been proposed to this end differ only in their choice of hydrophobicity scale.

The first, and still most popular, method is due to Hopp and Woods (1981). They used a slightly modified Nozaki–Tanford scale (the most important difference being that Asp, Glu, Arg, and Lys were all given equal hydrophobicity values) (Table 5.5), and scanned the sequence with a 6-residue window, concluding that the segment of maximal hydrophilicity was a good candidate for an antigenic site. The second and third highest points did not give reliable predictions. Parker *et al.* (1986) have used their scale based on HPLC retention times (Table 5.4), and a 7-residue window to make the same kind of prediction.

TABLE 5.5. **Antigenicity Scales**[a]

Residue	Hopp–Woods	Welling	Protrusion index *(PI)*	B_0	B_1	B_2
Arg	3.0	0.058	4.67	1.038	1.028	0.901
Asp	3.0	0.065	5.77	1.033	1.089	0.932
Glu	3.0	−0.071	5.43	1.094	1.036	0.933
Lys	3.0	0.206	5.57	1.093	1.082	1.057
Ser	0.3	−0.026	5.65	1.169	1.048	0.923
Asn	0.2	−0.077	4.82	1.117	1.006	0.930
Gln	0.2	−0.011	4.95	1.165	1.028	0.885
Gly	0.0	−0.184	5.31	1.142	1.042	0.923
Pro	0.0	−0.053	5.37	1.055	1.085	0.932
Thr	−0.4	−0.045	4.75	1.073	1.051	0.934
Ala	−0.5	0.115	4.55	1.041	0.946	0.892
His	−0.5	0.312	3.17	0.982	0.952	0.894
Cys	−1.0	−0.120	2.75	0.960	0.878	0.925
Met	−1.3	−0.385	2.83	0.947	0.862	0.804
Val	−1.5	−0.013	3.17	0.982	0.927	0.913
Ile	−1.8	−0.292	2.71	1.002	0.892	0.872
Leu	−1.8	0.075	2.93	0.967	0.961	0.921
Tyr	−2.3	0.013	3.22	0.961	0.930	0.837
Phe	−2.5	−0.141	2.61	0.930	0.912	0.914
Trp	−3.4	−0.114	2.70	0.925	0.917	0.803

[a]Note that the *B* index is different depending on whether the residue is surrounded by zero (B_0), one (B_1), or two (B_2) rigid residues (see text).

Welling *et al.* (1985) derived a statistical scale, where the antigenicity value for each kind of residue was calculated as the log of the quotient between its percentage in a sample of known antigenic regions and its percentage in average protein. They then analyze their sequences by averaging this value over a 7-residue moving window.

Other statistical scales are the protrusion index *(PI)* of Thorton *et al.* (1985), which measures the tendency of each kind of residue to be located in protruding portions of the chain, and the flexibility index (*B* index) of Karplus and Schultz (1985) based on crystallographic temperature factors. In the latter method, nearest-neighbor effects are taken into account by assigning different flexibility values to a given kind of residue depending on whether it is surrounded by 0, 1, or 2 "rigid" residues (i.e., Ala, Leu, His, Val, Tyr, Ile, Phe, Cys, Trp, and Met).

The reliability of these methods is hard to assess at present. Tanaka *et al.* (1985) synthesized 35 peptides, chosen on the basis of Hopp–Woods analyses of six oncoproteins, and tried to raise antibodies to these in rabbit. Thirty-two of the peptides (91%) elicited antipeptide antibodies, but only 18 of these antibodies (56%) reacted with their respective oncoprotein. Length appeared to be important for the effectiveness of the peptides, since all peptides with fewer than 10 residues failed to induce antibody response. The Hopp–Woods score and the position of the peptide in the protein did not correlate with the effect registered, however; indeed, the three highest-scoring peptides all failed to give antibodies that reacted with the respective protein.

E. Predicting the Structure of Membrane Proteins

Hydrophobicity analysis, i.e., hydrophobicity profiles and hydrophobic moment plots, is central to current methods for predicting (or, perhaps more correctly, for making informed guesses about) the structure of membrane proteins. Since such proteins are notoriously difficult to crystallize and hence to study by X-ray diffraction, prediction methods based on sequence analysis have become even more important than for soluble proteins. Fortunately, sequences of membrane proteins usually show a fairly clear-cut pattern of long, highly hydrophobic segments, separated by more polar regions with an amino acid composition characteristic of globular proteins. Thus, those parts of the chain that are embedded in the membrane most likely form long, helical, membrane-spanning rods that should be a lot more easy to predict with a reasonable degree of accuracy than the intricate weavings of the chain in a globular protein. In fact, the classical secondary structure prediction methods developed for globular proteins (Section II) perform exceptionally poorly on membrane proteins (Wallace *et al.*, 1986).

A second kind of membrane-binding structure is the amphiphilic helix, i.e., rather long helical segments with one markedly hydrophobic face, one charged or polar face, and a high hydrophobic moment (Kaiser and Kezdy, 1984). Such helices bind parallel to the surfaces of membranes and sometimes have lytic properties. Obviously, amphiphilic helices show up as high peaks in a hydrophobic moment analysis (Section V,C).

The most widely used method to predict the structure of membrane proteins is that of Kyte and Doolittle (1982): It is conceptually simple, it is fast, and it is easy to program. By comparing a set of membrane proteins and a set of soluble proteins, these authors found that the best discrimination between the two sets was obtained with a 19-residue window and the requirement that the mean hydrophobicity (on the Kyte–Doolittle Scale, Table 5.4) should be higher than +1.6 for predicting a membrane-spanning segment. Thus, such segments can be readily identified on a hydrophobicity plot. Similar schemes have been developed by others, e.g., Engelman *et al.* (1986), Argos *et al.* (1982), Klein *et al.* (1985), and Kuhn and Leigh (1985), but the Kyte–Doolittle method has, for better or worse, become the routine choice.

Engelman *et al.* (1986) have scrutinized the various schemes and scales, and provide the following guidelines: The window length should be around 20 residues; the Goldman–Engelman–Steitz (GES) scale (Table 5.4) may be the most appropriate for finding transmembrane helices; and pairs of oppositely charged residues at positions $i, i+3$ and $i, i+4$ (i.e., neighbors on the helical face) are stabilized by ~ 10 kcal/mol relative to the isolated residues. They contend that "a suitable scale and protocol can lead to the successful identification of transmembrane helical structures in integral membrane proteins."

Trying to assess our current predictive abilities, I recently analyzed a large collection of bacterial inner membrane proteins (von Heijne, 1986e). From the results, it was clear that no sharp line could be drawn between segments that are transmembrane and those that are not. There is a "no-man's land" where the mean hydrophobicity per residue for a 19-residue window is between 0.9 and 1.4 kcal/mol (GES scale); segments that fall in this region cannot be assigned unambiguously as transmembrane or solvent exposed. A second result was that the amino acid composition of the loops connecting the transmembrane segments correlates with the transmembrane topology of the molecule: Loops on the cytoplasmic face of the membrane are about fourfold enriched in Arg and Lys residues as compared with periplasmic loops. Thus, to the extent that the transmembrane segments can be identified, the orientation of the protein relative to the membrane can also be predicted.

As a straightforward extension of the Kyte–Doolittle type of analysis, the hydrophobic moment can also be incorporated into the prediction. In membrane proteins with many membrane-spanning segments, helices with high hydrophobic moments may be assumed to have their apolar face exposed to the surrounding lipid and their polar face interacting with other membrane-embedded parts of the protein, or lining transmembrane channels. Surface-binding segments in peripheral membrane proteins can also be identified, e.g., on plots of $<\mu>$, the mean hydrophobic moment per residue, versus $<H>$, the mean hydrophobicity of tentative membrane-interacting segments (Fig. 5.3) (Eisenberg *et al.,* 1984a).

The Eisenberg prediction scheme is interesting also from another point of view, namely that it identifies membrane-spanning segments in two steps: First, all candidates with $<H> \geq 0.42$ [Eisenberg consensus scale (Table 5.4) with a 21-residue window] are picked out; second, these candidates are kept only if (a) there is at least one candidate with $<H> \geq 0.68$ in the protein or (b) there are at least two candidates whose summed $<H>$ values are ≥ 1.10. In this way, some globular proteins with only one candidate segment of intermediate $<H>$ are eliminated.

Schemes like this have been applied extensively to channel-forming proteins such as the acetylcholine receptor (Finer-Moore and Stroud, 1984; Guy, 1984), the sodium channel (Guy and Seetharamulu, 1986), and colicins (Pattus *et al.,* 1985), but it is not clear at present how far one can trust such detailed predictions.

Another generalization of the hydrophobicity and hydrophobic moment calculations is to let the window length vary and thus obtain an optimal segment length such that the total hydrophobicity is maximized (von Heijne, 1986a). This requires a hydrophobicity scale with an appropriate zero-point; the composition of known membrane-spanning segments indicates that the zero-point should be close to the value for Ala. In this way, the results can be presented on a two-dimensional plot with the optimal segment length versus its total hydrophobicity or hydrophobic moment (see Fig. 5.4 in Section VI,A below).

Finally, as has been pointed out by, e.g., Paul and Rosenbusch (1985), for proteins with only short loops protruding out of the membrane, turn-predictions analogous to those used in the classical secondary structure predictions may provide an alternative way of delineating putative membrane-spanning segments. These workers, as well as Vogel and Jähnig (1986), have also tried to predict the structures of a couple of proteins from the outer membrane of *E. coli* known to be predominantly β-structure. In this case, no long apolar stretches are seen, and the hydrophobic moment calculated with $\delta = 180°$ (alternating β-strand, Section V,C) was used to spot likely transmembrane β-strands.

At the present state of the art, however, it is clear that there is no

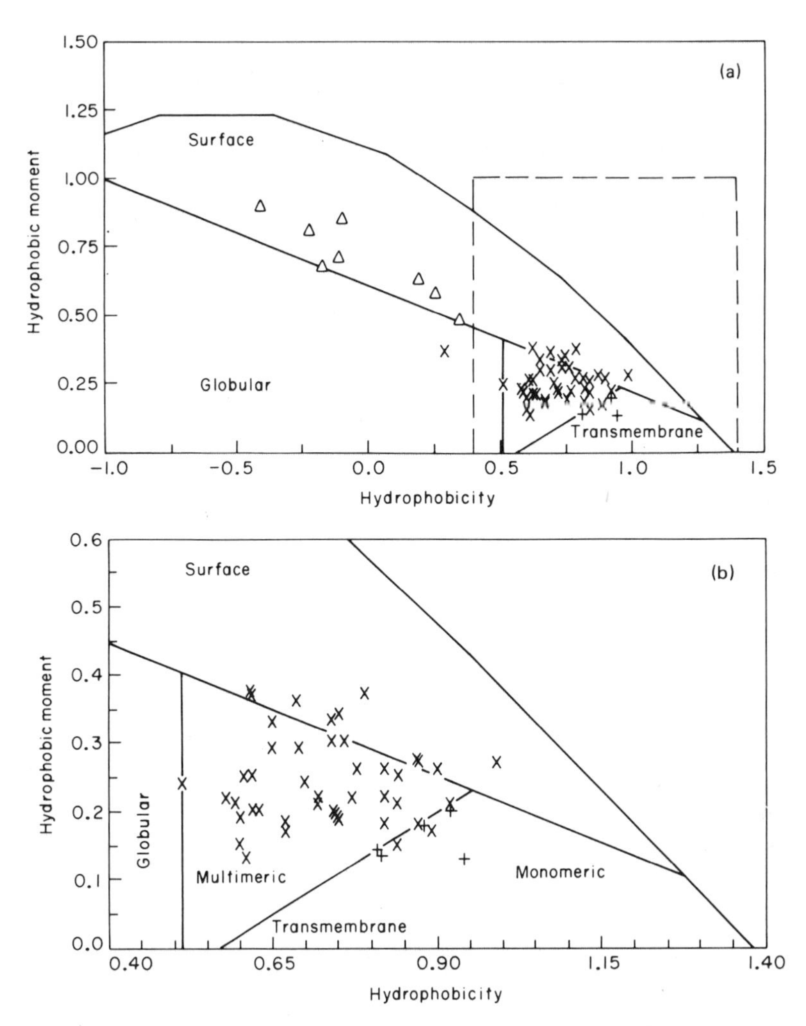

Fig. 5.3. Plot of $<\mu>$ versus $<H>$ (averages over 11 residues) for a number of surface-associated (Δ, panel a) and membrane-spanning protein segments (panel b). Note that membrane-spanning segments from proteins with multiple transmembrane parts (multimeric, $\times$) tend to have lower $<H>$ and higher $<\mu>$ than spanning segments from proteins that cross the membrane only once (monomeric, $+$), i.e., the latter have a more uniformly apolar helical surface. (From Eisenberg *et al.*, 1984a.)

rough-and-ready method that will give the "true" structure at the push of a button. As an illustration, consider the predictions listed in Table 5.6, where the Kyte–Doolittle and Eisenberg methods have both been applied to a set of mitochondrial membrane proteins. Unless an addicted gambler, one would not like to put down too much in a bet on one of these predictions.

VI. SORTING SIGNALS IN PROTEINS

In prokaryotic and eukaryotic cells alike, the quasitotality of all proteins are made in one compartment, the cytoplasm. During or after synthesis, some of these proteins are sorted according to their final destination in the cell: Proteins are secreted from the cell; imported into mitochondria, chloroplasts, microbodies (peroxisomes, glyoxysomes, glycosomes), and the nucleus; or inserted into one or other target membrane (the endoplasmic reticulum, the plasma membrane,

***TABLE 5.6.* Number of Membrane Segments Predicted by the Kyte–Doolittle and Eisenberg Schemes for a Collection of Mitochondrial Inner-Membrane Proteins**[a]

	No. of predicted membrane segments	
Protein	Kyte–Doolittle	Eisenberg
Cytochrome *c* oxidase I	8	12
Cytochrome *c* oxidase II	2	3
Cytochrome *c* oxidase III	3	6
Cytochrome *b*	6	10
ATPase 6	6	6
URF1	6	9
URF2	4	9
URF3	3	3
URF4	4	10
URF4L	2	3
URF5	14	15
URF6	4	4
URFA6L	1	1

[a]From von Heijne (1986c).

the outer and inner membrane of bacteria, mitochondria, and chloroplasts). Sorting of this kind obviously involves transport across one or more cellular membrane, and seems to require a sorting signal on the protein as well as specific receptors on the appropriate membranes (see Wickner and Lodish, 1985, for a general review).

All experimental findings agree that the sorting signals reside in the amino acid sequence, either in the form of short, contiguous segments, or in the form of some specific three-dimensional structure involving residues that are distant in the chain. Some of these signals have been rather well characterized—secretory signal peptides (SPs) and mitochondrial and chloroplast targeting or transit peptides (TPs)—whereas others are less well defined. In this section, I will present a couple of pattern-recognition schemes designed to find particular classes of sorting signals.

A. Secretory Signal Peptides

SPs target proteins to the secretory pathway. In prokaryotes, this includes the inner and outer membranes, the periplasmic space, and the extracellular medium. In higher cells, the secretory pathway goes from the endoplasmic reticulum (ER), through Golgi and post-Golgi compartments, to the plasma membrane and the exterior of the cell; a subsidiary route ends in the lysosome. Regardless of the final destination, the initial event is always SP-mediated secretion across the ER membrane (Rapoport, 1986).

Statistical and experimental studies of a large number of known SPs have provided an approximate picture of the minimal SP (von Heijne, 1985): a short, positively charged amino-terminal region (*n* region) immediately following the initiator Met, a 7- to15-residues-long central hydrophobic core (*h* region), and a more polar 3- to 7-residues long carboxy-terminal region (*c* region) leading up to the site of cleavage between the SP and the mature protein. The general outline of an SP is thus similar to that of transmembrane segments in integral membrane proteins, but the *h* region is often shorter (Fig. 5.4). The only reasonably well-conserved region is the *c* region, and acceptable cleavage sites seem to follow a (−3, −1)-rule with small, uncharged residues in positions −1 and −3 counting from the cleavage site.

Two procedures for automatically identifying SPs have been published. McGeoch (1985) requires that an acceptable SP should have an *n* region of length ≤ 11 residues and net charge between −1 and +2,

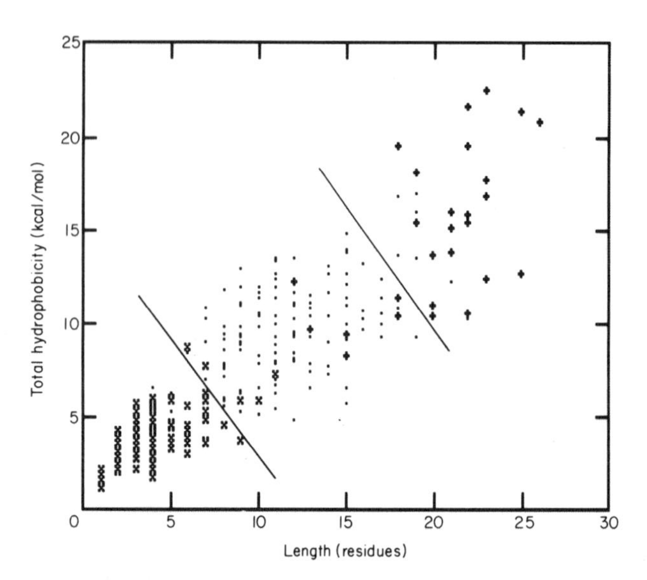

Fig. 5.4. Plot of total hydrophobicity versus length of the most hydrophobic segments in samples of 134 eukaryotic cytoplasmic proteins (40 amino-terminal residues, x), 170 eukaryotic signal peptides (.), and 30 eukaryotic membrane-spanning segments (+). A renormalized GES scale (Table 5.4) with the value for Ala = 0.0 was used, and the segment of highest summed hydrophobicity, not normalized on a per residue basis, was located in each sequence. (From von Heijne, 1986a.)

and an *h* region of length ≥ 8 residues where the total hydrophobicity of the maximally hydrophobic 8-residue stretch is roughly ≥ 15 on the Kyte-Doolittle scale. These criteria resolve some 95% of a sample of 114 eukaryotic SPs from a sample of cytoplasmic proteins.

I have developed a method that both finds likely prokaryotic and eukaryotic SPs and predicts the most likely site of cleavage (von Heijne, 1986b). The method is based on a standard weight matrix approach (Chapter 4, Section II,A,2). Tables 5.7 and 5.8 show amino acid counts in samples of 36 prokaryotic and 161 eukaryotic SPs aligned from their site of cleavage. The actual weight matrices are obtained by first dividing all counts by their respective expected abundance in soluble proteins in general, and then taking the natural logarithms of these quotients. All zero counts are put equal to one before the division, except for positions −3 and −1, where zero counts are put equal to some arbitrarily small value (10^{-10}) before taking the logs. The most probable cleavage site is then found by scanning the se-

TABLE 5.7. Amino Acid Counts for 161 Eukaryotic Signal Peptides[a]

Residue	−13	−12	−11	−10	−9	−8	−7	−6	−5	−4	−3	−2	−1	+1	+2	Expected
A	16	13	14	15	20	18	18	17	25	15	47	6	80	18	6	14.5
C	3	6	9	7	9	14	6	8	5	6	19	3	9	8	3	4.5
D	0	0	0	0	0	0	0	0	5	3	0	5	0	10	11	8.9
E	0	0	0	1	0	0	0	0	3	7	0	7	0	13	14	10.0
F	13	9	11	11	6	7	18	13	4	5	0	13	0	6	4	5.6
G	4	4	3	6	3	13	3	2	19	34	5	7	39	10	7	12.1
H	0	0	0	0	0	1	1	0	5	0	0	6	0	4	2	3.4
I	15	15	8	6	11	5	4	8	5	1	10	5	0	8	7	7.4
K	0	0	0	1	0	0	1	0	0	4	0	2	0	11	9	11.3
L	71	68	72	79	78	45	64	49	10	23	8	20	1	8	4	12.1
M	0	3	7	4	1	6	2	2	0	0	0	1	0	1	2	2.7
N	0	1	0	1	1	0	0	0	3	3	0	10	0	4	7	7.1
P	2	0	2	0	0	4	1	8	20	14	0	1	3	0	22	7.4
Q	0	0	0	1	0	6	1	0	10	8	0	18	3	19	10	6.3
R	2	0	0	0	0	1	0	0	7	4	0	15	0	12	9	7.6
S	9	3	8	6	13	10	15	16	26	11	23	17	20	15	10	11.4
T	2	10	5	4	5	13	7	7	12	6	17	8	6	3	10	9.7
V	20	25	15	18	13	15	11	27	0	12	32	3	0	8	17	11.1
W	4	3	3	1	1	2	6	3	1	3	0	9	0	2	0	1.8
Y	0	1	4	0	0	1	3	1	1	2	0	5	0	1	7	5.6

[a]The cleavage site is between positions −1 and +1. (From von Heijne, 1986b.)

TABLE 5.8. **Amino Acid Counts for 36 Prokaryotic Signal Peptides**[a]

Residue	−13	−12	−11	−10	−9	−8	−7	−6	−5	−4	−3	−2	−1	+1	+2	Expected
A	10	8	8	9	6	7	5	6	7	7	24	2	31	18	4	3.2
C	1	0	0	1	1	0	0	1	1	0	0	0	0	0	0	1.0
D	0	0	0	0	0	0	0	0	0	0	0	0	0	2	8	2.0
E	0	0	0	0	0	0	0	0	0	0	0	1	0	4	8	2.2
F	2	4	3	4	1	1	8	0	4	1	0	7	0	1	0	1.3
G	4	2	2	2	3	5	2	4	2	2	0	2	2	1	0	2.7
H	0	0	1	0	0	0	0	1	1	0	0	7	0	1	0	0.8
I	3	1	5	1	5	0	1	3	0	0	0	0	0	0	2	1.7
K	0	0	0	0	0	0	0	0	0	1	0	2	0	3	0	2.5
L	8	11	9	8	9	13	1	0	2	2	1	2	0	0	1	2.7
M	0	2	1	1	3	2	3	0	1	2	0	4	0	0	1	0.6
N	0	0	0	0	0	0	0	1	1	1	0	3	0	1	4	1.6
P	0	1	1	1	1	1	2	3	5	2	0	0	0	0	5	1.7
Q	0	0	0	0	0	0	0	0	2	2	0	3	0	0	1	1.4
R	0	0	0	0	0	0	0	0	0	0	0	0	0	1	0	1.7
S	1	0	1	4	4	1	5	15	5	8	5	2	2	0	0	2.6
T	2	0	4	2	2	2	2	2	5	1	3	0	1	1	2	2.2
V	5	7	1	3	1	4	7	0	0	4	3	0	0	2	0	2.5
W	0	0	0	0	0	0	0	0	0	0	0	0	0	1	0	0.4
Y	0	0	0	0	0	0	0	0	0	3	0	1	0	0	0	1.3

[a] The cleavage site is between positions −1 and +1. (From von Heijne, 1986b.)

quence with the appropriate prokaryotic or eukaryotic weight matrix, summing the weights for each position of the matrix, and taking the highest-scoring position to be the most likely processing site.

This method is estimated to predict the correct site of cleavage 75–80% of the time when applied to SPs outside the data base, and to resolve 95% of all SPs (highest score > 3.5) from 95% of all non-SP amino-termini (highest score < 3.5). It is not foolproof, however, and will not in all cases correctly classify nonfunctional SPs resulting from, e.g., point mutations; these are still sufficiently different from the amino-termini of normal cytoplasmic proteins to sometimes score as SPs. The weight matrix method when complemented by McGeoch's method and/or the rules for acceptable minimal SPs should be able to weed out even such pathological SPs, and achieve a very high reliability.

B. Mitochondrial and Chloroplast Transit Peptides

In most cases, nuclearly encoded mitochondrial and chloroplast proteins are made with transient amino-terminal transit peptides (TPs) and are imported posttranslationally (Douglas *et al.,* 1986; Schmidt and Mishkind, 1986). Theoretical and experimental analyses of TPs are at present less conclusive than for SPs: Besides a marked preponderance of positively charged residues and a near absence of negatively charged ones, the only possibly common feature of mitochondrial TSs noted so far is a potential to fold into reasonably strong amphiphilic helices (von Heijne, 1986d). Chloroplast TSs also have few if any negatively charged residues, and at least some of them are thought to share a common framework of limited amino acid similarity (Karlin–Neumann and Tobin, 1986).

Suborganellar sorting within the mitochondrion and the chloroplast also seems to be determined by signals in the amino acid sequence, such as apolar transmembrane stretches or other membrane-interacting segments (Douglas *et al.,* 1986; Smeekens *et al.,* 1986).

No *bona fide* methods designed to find mitochondrial or chloroplast TSs exist so far.

C. Protein Import into Other Organelles

Protein import into microbodies is even less well understood, and does not seem to require cleavable sorting signals. The only difference between microbody enzymes and their cytoplasmic isozymes noted thus far is a much higher positive net charge in the imported protein, but the significance of this observation is unknown (Borst, 1986). One possibility is that critically spaced positively charged residues serve as import signals (Wierenga *et al.,* 1987).

Nuclear import is a little better understood, and a few short nuclear uptake peptides (NPs) have been defined by gene fusion and deletion studies. A. E. Smith *et al.* (1985) have shown that the sequence Pro-Pro-Lys-Lys-Lys-Arg-Lys-Val from the Simian virus 40 (SV40) large T antigen can direct foreign proteins into the nucleus. The first 21 residues of the yeast ribosomal protein L3, which includes a sequence Pro-Arg-Lys-Arg, can also bring fused proteins into the nucleus (Moreland *et al.,* 1985). Other signals may also be important (Richardson *et al.,* 1986; Davey *et al.,* 1985). A. E. Smith *et al.* (1985) did attempt to construct a consensus sequence based on the SV40 NP, but had little success when scanning the NBRF data bank with this sequence.

VII. COVALENT MODIFICATIONS OF PROTEINS

In addition to the removal of sorting signals and activation peptides, many proteins undergo various covalent modifications such as glycosylation and acetylation. In a sense, degradation through the action of intracellular proteases represents the ultimate modification. Experimental and statistical analyses have shed some light on the sequence determinants involved, and some predictions as to the fate of a given protein can be made from its amino acid sequence alone.

A. Glycosylation

Most proteins that pass through the secretory pathway in eukaryotic cells are modified by the addition of oligosaccharides to Asn residues. The oligosaccharides on the mature proteins are of two kinds, high-mannose and complex, but both kinds come from a common precursor, a large, high-mannose oligosaccharide that is transferred *en bloc* from the carrier lipid dolichol phosphate present in the ER membrane. While the protein is en route to its final location, the high-mannose moieties can be modified in various ways by addition of sugars such as galactose, fucose, and sialic acid (see Hubbard and Ivatt, 1981, for a review).

The high-mannose precursor is always transferred to Asn residues that have Ser or Thr 2 residues to their carboxy-terminal side: the Asn-X-(Thr-Ser) tripeptide (where X is any amino acid) is the universal acceptor. Interestingly, Asn-X-Thr and Asn-X-Ser tripeptides are much less abundant in secretory eukaryotic proteins than what one would expect from the overall frequencies of Asn, Thr, and Ser residues—they are found at only 40% of the expected abundance—whereas no such reduction is seen in either intracellular eukaryotic proteins or in any group of prokaryotic proteins (Hunt and Dayhoff, 1970; Sinohara and Maruyama, 1973). Nevertheless, only about 30% of the putative glycosylation sites are in fact modified, and those that are are probably exposed on the surface of the proteins (Hubbard and Ivatt, 1981). It has also been observed that oligosaccharides of the complex kind are often found closer to the amino-terminus than the high-mannose groups (Pollack and Atkinson, 1983).

From the point of view of trying to predict which sites are likely to be glycosylated and which are not, most attempts have so far tried to correlate glycosylation with predicted turns (e.g., Aubert *et al.,* 1976); considering the less-than-perfect secondary structure prediction methods available (Section II), it comes as no surprise that this has not proven to be a very reliable guide.

B. Amino-Terminal Modifications (Met Removal and Acetylation)

During maturation, the amino termini of both intracellular and extracellular proteins undergo a number of modifications such as deformylation (in prokaryotes) or removal of the initiator Met, and/or acetylation. A statistical analysis of a large sample of proteins has revealed that the amino-terminal methionine aminopeptidase of both prokaryotic and eukaryotic cells apparently has a rather strict amino acid specificity (Flinta *et al.,* 1986): Lys, Arg, Leu, and (in prokaryotes) Phe and Ile do not allow removal of the amino-terminal Met when next to it in the sequence; Ala, Gly, Pro, Ser, Thr, and (in eukaryotes) Val promote its removal. Residues that have no clear-cut effect on Met removal rarely appear next to it in wild-type proteins. This pattern has also been found in experimental studies (Tsunasawa *et al.,* 1985).

Incidentally, the preference for Ala, Gly, and Val next to the initiator Met in eukaryotes may in part explain Kozak's rule (Chapter 4, Section V,B,2), namely, that ribosome-binding sites in eukaryotic mRNAs have a consensus ANN*AUG*G: These three amino acids all cause Met removal, and all have codons starting with G. The 5′ G in the consensus may thus at least to some extent be a result of selection acting on the protein sequence rather than on the mRNA.

Amino-terminal acetylation is even more restricted. In a sample of acetylated proteins, Ala, Ser, and Met were found to be the amino-terminal residue in more than 85% of the cases (Persson *et al.,* 1985). Acetylated Met residues were invariably followed by Asp or Glu, but cytochrome *c* mutants with other penultimate residues following an acetylated Met have also been observed (Tsunasawa *et al.,* 1985).

Met removal can be quite reliably predicted from these observations, since 80% of the sequences analyzed by Flinta *et al.* (1986) had penultimate residues falling in one of the two groups with clear-cut effects. Acetylation is less easily predicted, since residues beyond the amino-terminal one apparently have an influence (Tsunasawa *et al.,* 1985). Met-Asp pairs or Ser at the amino terminus seem to be the best indicators of acetylation at present.

C. Protein Degradation

In vivo, proteins have widely different half-lives ranging from many hours down to minutes. The factors responsible for these differences have remained largely unknown until recently, but it now seems that the rate of turnover can be controlled by the identity of the amino-terminal group or amino acid, at least for proteins degraded through the ubiquitin pathway.

First, Hersko *et al.* (1984) could show that amino-terminal acetylation prevented ubiquitin-mediated degradation—this of course implicitly depends on the amino-terminal residue as described above—and very recently Bachmair *et al.* (1986) have used an elegant approach based on gene fusions to demonstrate that certain amino-terminal residues are largely resistant to ubiquitin whereas others are extremely sensitive. Surprisingly, the resistant residues are precisely those that were shown above to promote Met removal. In addition, Met itself is also resistant, whereas all other residues, notably those that do not promote Met removal, are highly sensitive. Since almost all normal proteins will have either Met or one of the residues promoting Met removal at their amino terminus, they will all probably be resistant, until they are cleaved or damaged at some internal position, thus exposing a ubiquitin-sensitive residue.

This last point has been addressed by Rogers *et al.* (1986), who have found regions rich in Pro, Glu, Ser, and Thr (PEST in the one-letter notation) in a number of short-lived proteins but only in a few stable ones. They have constructed a PEST score that is a composite of the percent PEST residues and the mean Kyte–Doolittle hydrophobicity index in regions flanked by Arg, Lys, or His residues, and suggest this as a means of predicting the half-life of a protein. The reliability of this score is hard to estimate, and should be evaluated on samples of stable and unstable proteins not included among the original ones.

Chapter 6

Sequence Similarities, Homologies, and Alignments

If you have done nothing else in terms of theoretical sequence analysis, no doubt you have more than once asked a colleague to run your latest sequence through a homology-search program in the hope that it will turn out to be related to some interesting protein already stored in one of the data banks. And your colleague more likely than not has come back with a list of more or less surprising, more or less convincing "matches," a stack of dot-matrix comparisons, and one or two alignments with the best candidates.

In this chapter, I will describe a number of different methods that are presently used in this area. Following Davison (1985), three main categories of sequence comparison will be dealt with (leaving his fourth category, visual comparison, to the reader's imagination): dot-matrix plots, global sequence alignments, and local sequence alignments. I will also briefly discuss various ways to estimate the statistical significance of a given match.

A word on terminology: People in the field shudder when the terms "similarity" and "homology" are used indiscriminately: Similarity simply means that sequences are in some sense similar and has no evolutionary connotations, whereas homology refers to evolutionarily related sequences stemming from a common ancestor.

I. DOT-MATRIX ANALYSIS

Davison (1985) states that, "The use of matrix comparisons is very highly recommended for sequences of all sizes, since significant features can be overlooked by other methods." In addition, the basic dot-matrix plot routine is very easy to program even on a small personal computer. The main difficulty with this method is to estimate the significance of the patterns one observes: "Precise identification of the matched sequences requires some effort, and the statistical significance of the matches is not immediately obvious, but the overall picture of the 'forest' is communicated" (Pustell and Kafatos, 1982).

We have encountered dot matrices already in Chapter 4, section VII,A (RNA folding). Briefly, one constructs a matrix with one of the sequences to be compared running across the top, and the other running down the left-hand side. In each element in the matrix, one enters a measure of the similarity of the two residues corresponding to that element. The simplest scoring system distinguishes only between identical (dots) and nonidentical (blanks) residues, but one can also use graded measures that give chemically similar amino acids high similarity scores (Staden, 1982a; see Section V below). Similar sequences tend to have many identical or chemically related residues along the main diagonal; hence conspicuous diagonal runs of dots signal regions of similarity (or homology, if the evolutionary argument can be carried through).

One of the first published accounts of the dot-matrix method is by Gibbs and McIntyre (1970), who placed dots at the positions of identical residues. They also presented two ways to assess the significance of the diagonal runs: first, by comparing the lengths of the longest unbroken runs with the lengths expected from a comparison of random sequences; second, by calculating the total number of matches (dots) in the principal and adjacent diagonals, and comparing with the random expectation.

The utility of the dot-matrix method can be greatly enhanced by a variety of filtering techniques. This is especially important when one is comparing nucleic acid sequences, since around 25% of the matrix will be filled with dots even for random sequences. Maizel and Lenk (1981) and Staden (1982a) describe programs that can filter the data by, e.g., only plotting those dots that belong to diagonal runs of a certain minimal length, those that belong to runs with at least m matches out of n consecutive residues, or those that belong to runs that are likely not to have appeared by chance as judged by some statistical model. Pustell and Kafatos (1982) use an exponential damping function in their calculation of the score for each match, i.e., they compute a homology score in which the contributions of progressively more distant matches are given relatively less weight in proportion to their distance from the aligned base pair; random local matches and mismatches are well suppressed.

Programs such as these are very fast, and can be run interactively: One can first make a comparison at high stringency to get the overall picture and then progressively focus on smaller regions using less stringent filters. These programs also often incorporate compression algorithms, whereby matrices representing sequences many thousands of residues long can be compressed onto a single page.

II. GLOBAL ALIGNMENTS

An optimal global alignment strikes a balance between the total number of matches (identities or similarities) in the alignment and the number of gaps (insertions or deletions) used to obtain the result. By assigning weights to each kind of match (or mismatch) and each gap length, a total score can be calculated for any conceivable alignment and the optimal one(s) can in principle be found simply by repeating the calculation for all possible cases.

As so often happens, the combinatorics of such a brute force approach quickly become unmanageable. Fortunately, methods exist that will generate at least one of the optimal solutions in a reasonable time with reasonable memory requirements. These methods belong to the category of dynamic programming algorithms mentioned in the section on RNA folding (Chapter 4, Section VII,A); here, I will briefly review the underlying idea and some rules of thumb that are important to keep in mind when using programs of this sort. More comprehensive reviews can be found in, e.g., Davison (1985) and Goad (1986).

The basic method was first described by Needleman and Wunsch (1970), and later in a slightly different form by Sellers (1974). As in the RNA-folding scheme discussed above, one constructs two matrices: a distance matrix **D** and a traceback matrix **T**. Each element D_{ij} will eventually contain a measure of the minimal distance between subsequences $A_1 \ldots A_i$ and $B_1 \ldots B_j$ in the optimal alignment, and the information in the element T_{ij} will make it possible to recreate the path through **D** that defines this alignment.

Before one starts, one must assign weights to gaps and mismatches: If one is simply scoring identities, a match will have weight $\beta = 0$, and a mismatch will have $\beta = 1$ (one can also use some similarity scale, see

Section V). The most frequently used gap-weighting function is of the form

$$W(k) = a + bk \tag{6.1}$$

where a and b are constants, and k is the gap length.

In Sellers' method, the two sequences to be aligned, A and B, are written across the top and down the left-hand side of **D** and **T**, and **D** is initialized by putting $D_{0,0} = 0$ and $D_{0,k} = D_{k,0} = W(k)$(i.e., a misalignment at the end is counted as a gap). The elements in **D** are then generated recursively by the relation

$$D_{i,j} = \min \begin{cases} D_{i-1,j-1} + \beta(a_i,b_j) \\ D_{i,j-k} + W(k);\ k=1,..,j-1 \\ D_{i-k,j} + W(k);\ k=1,..,i-1 \end{cases} \tag{6.2}$$

The first term represents a continuation from the previous diagonal element, adding the degree of mismatch between a_i and b_j; the second term represents all possible continuations by making gaps from the previously calculated horizontal elements; and the third term represents all continuations with gaps from the previously calculated vertical elements. The value finally entered in $D_{i,j}$ is the smallest of the three terms. When this has been found, $T_{i,j}$ is put equal to zero if the optimal continuation was from $D_{i-1,j-1}$; if it was from a horizontal or vertical element the corresponding k value is entered (with negative sign in the latter case, to distinguish the two alternatives).

When the whole **D** matrix has been generated in this way, the smallest distance separating the two sequences is found in the final row or column, and the corresponding optimal alignment is reconstructed by moving back through the **T** matrix and inserting the appropriate gaps.

More often than not, more than one alignment will have the same minimal distance; the algorithm outlined here will only find one of these. Of course, it will not find any "near-optimal" alignments, either. Procedures that do find all optimal as well as near-optimal alignments have been published (Beyers and Waterman, 1984; Boswell and McLachlan, 1984), but are not in general use.

It is important to realize that an optimal alignment is optimal only for the particular values chosen for the mismatch and gap weights. When any of these is altered, the optimal alignment will also change. Fitch and Smith (1983) have demonstrated how sensitive an alignment can be to small changes in the parameters, and have summarized their results in the form of a number of rules of thumb (Table 6.1). Even with

TABLE 6.1. **Rules for Sequence Similarity Searching**[a]

A. Methodological
1. The alignment procedure must provide weights for a gap *(a)* and for the length of the gap *(b)*.
2. a and b must never be zero.
3. The sum of $a + b$ must be greater than the weight β for a single mismatch if insertions and deletions are expected to occur less frequently than substitutions.
4. If there are two strongly matched regions separated by a region of low similarity, the weight for a gap a is the most important parameter to explore for determining the optimal alignment, since a affects the number of gaps in the region of low similarity.
5. If the two sequences have no obvious relationship at their right and left ends, gaps at the ends should not be penalized. The Needleman–Wunsch algorithm is preferable in this case.
6. Weight for mismatches other than one should be permitted; a transition may not be as unlikely as a transversion. In the case of amino acids, an Ile-Val change should be less penalized than an Ile-Arg change, etc.
7. One must make some estimate of the probability that a given alignment is due to change.

B. Interpretive
1. A gap and its length are distinct quantities. Cf. a and b in expression (6.1) for $W(k)$.
2. An optimal alignment is by no means necessarily statistically significant.
3. One searches for similarity; homology is an evolutionary inference based on examination of the similarity and its biological meaning. Similarity, homology, and analogy have distinct, and different, meanings in evolutionary biology. Thus, a similarity may result from a homology or an analogy, but not both.

[a]From Davison (1985). Reprinted with permission from Bulletin of Mathematical Biology, Vol. 47. Copyright 1985, Pergamon Journals, Ltd.

the very best of programs at one's disposal, it still requires some degree of experience to draw the right conclusions from the results produced —and a good grasp of the biology of the problem at hand is as essential as one's familiarity with computers.

Although most often implemented on mainframe computers, algorithms of this kind can in fact also be programmed on personal computers. As an example, Tyson and Haley (1985) have written a program in PASCAL that runs on an Apple IIe computer equipped with a minimum of 48K of memory. In my own experience, this program works very well and is not overly slow.

III. LOCAL ALIGNMENTS

A second aspect of the similarity/alignment problem is to search for common subsequences in the proteins being compared. A global alignment can easily miss such segments. The Needleman–Wunsch–Sellers (NWS) method can still be used in a slightly modified form—one that guarantees that all subsequence matches above a certain threshold will be found. A completely different approach is to start out looking for short perfect identities, and build longer subsequence (and even global) alignments from these. Programs based on the latter method can be made to run extremely fast, and are presently used to scan entire data banks for matches to a given sequence.

A. NWS Algorithms

Smith and Waterman (1981) typifies the modified NWS method. It is very similar to the global alignment method described above (except that it happens to be formulated in terms of the maximum similarity rather than the minimum difference between the optimal subsequences). Since the register between the two sequences is of no importance (we will let them be free to slide relative to one another), the **D** matrix is initialized with all zeros in the first row and column, i.e., $D_{0,j} = D_{i,0} = 0$. Now, we calculate the elements of **D** such that $D_{i,j}$ is the maximum similarity of two segments ending at a_i and b_j in sequences A and B:

$$D_{i,j} = \max \begin{cases} D_{i-1,j-1} + \beta(a_i,b_j) \\ D_{i-k,j} - W(k);\ k=1,..,j-1 \\ D_{i,j-k} - W(k);\ k=1,..,i-1 \end{cases} \tag{6.3}$$

If $D_{i,j} < 0$, we put it equal to zero, indicating no similarity up to a_i and b_j.

An example from the original paper is shown in Fig. 6.1. This **D** matrix was calculated with $\beta = 1$ for an identical match, $\beta = -1/3$ for a mismatch, and $W(k) = 1 + k/3$. The underlined elements show the traceback from the largest element in **D**. Additional subsequence alignments can be obtained by starting from the second largest element, etc.

Goad and Kanehisa (1982) have developed a similar strategy. They construct two **D** matrices, one starting from the top left and working downward, one starting from the bottom right and working upward.

5′	5′													3′
		C	A	G	C	C	U	C	G	C	U	U	A	G
	0.0	0.0	0.0	0.0	0.0	0.0	0.0	0.0	0.0	0.0	0.0	0.0	0.0	0.0
A	0.0	0.0	1.0	0.0	0.0	0.0	0.0	0.0	0.0	0.0	0.0	0.0	1.0	0.0
A	0.0	0.0	1.0	0.7	0.0	0.0	0.0	0.0	0.0	0.0	0.0	0.0	1.0	0.7
U	0.0	0.0	0.0	0.7	0.3	0.0	1.0	0.0	0.0	0.0	1.0	1.0	0.0	0.7
G	0.0	0.0	0.0	1.0	0.3	0.0	0.0	0.7	1.0	0.0	0.0	0.7	0.7	1.0
C	0.0	1.0	0.0	0.0	2.0	1.3	0.3	1.0	0.3	2.0	0.7	0.3	0.3	0.3
C	0.0	1.0	0.7	0.0	1.0	3.0	1.7	1.3	1.0	1.3	1.7	0.3	0.0	0.0
A	0.0	0.0	2.0	0.7	0.3	1.7	2.7	1.3	1.0	0.7	1.0	1.3	1.3	0.0
U	0.0	0.0	0.7	1.7	0.3	1.3	2.7	2.3	1.0	0.7	1.7	2.0	1.0	1.0
U	0.0	0.0	0.3	0.3	1.3	1.0	2.3	2.3	2.0	0.7	1.7	2.7	1.7	1.0
G	0.0	0.0	0.0	1.3	0.0	1.0	1.0	2.0	3.3	2.0	1.7	1.3	2.3	2.7
A	0.0	0.0	1.0	0.0	1.0	0.3	0.7	0.7	2.0	3.0	1.7	1.3	2.3	2.0
C	0.0	1.0	0.0	0.7	1.0	2.0	0.7	1.7	1.7	3.0	2.7	1.3	1.0	2.0
G	0.0	0.0	0.7	1.0	0.3	0.7	1.7	0.3	2.7	1.7	2.7	2.3	1.0	2.0
G	0.0	0.0	0.0	1.7	0.7	0.3	0.3	1.3	1.3	2.3	1.3	2.3	2.0	2.0
3′														

```
                                    -G-C-C-A-U-U-G-
Optimal subsequence alignment:       ¦ ¦ ¦   ¦   ¦
                                    -G-C-C---U-C-G-
```

Fig. 6.1. **D** matrix for finding subsequence alignments in a short RNA sequence. The traceback from the largest element (= 3.3) is indicated by underlining. (From Smith and Waterman, 1981.)

The final matrix is obtained by including only those elements that are above the threshold (zero in the example above) in both **D** matrices. They further filter the subsequence alignments on the basis of an estimate of their statistical significance.

B. Hash-Coding Algorithms

Leaving the NWS methods aside for the moment, a second way to tackle the problem of finding local similarities is to use list-sorting or hashing routines. These are based on the construction of a list or "lookup table" of *k*-letter words or *k*-tuples (e.g., all possible di- or trinucleotides), and the positions where they appear in the sequences being compared. This method was first introduced into molecular biology by Dumas and Ninio (1982), and is now embodied in two of the most extensively used "fast-search" programs: the FASTP (for protein sequences) and FASTN (for nucleotide sequences) programs (Wilbur and Lipman, 1983; Lipman and Pearson, 1985). FASTP, since it is so much in use, will be described in some detail.

Consider the following amino acid sequences, *A*, GGHELVLQ, and *B*, PEVVLRS, where *A* is the query sequence and *B* is a sequence from the data bank. First, a lookup table for *A* is constructed (we choose one-letter words here, i.e., $k = 1$) by making a list with each type of amino acid on a separate line, and entering the position of each residue in *A* on the appropriate line. Thus, E appears at position 4; G at positions 1 and 2; H at position 3, etc. Then, for each residue in sequence *B*, one looks up the positions of all identical residues in the table for *A*, and calculates the difference between the position of the match in the two sequences (the offset). All matches having the same offset can be simultaneously aligned without intervening gaps. In the example, there are three matches with an offset of 2 (E_4 in *A* with E_2 in *B*, V_6 with V_4, and L_7 with L_5); all other offsets have less matches. Local regions of similarity can thus be identified from offsets with a high number of matches.

In FASTP, an initial similarity score is calculated for the five regions of highest similarity in the two sequences by using a so-called amino acid replaceability matrix (see Section V) that gives positive scores to evolutionarily frequent, conservative replacements of one amino acid for a chemically similar one. An optimized score can also be calculated in a second step by a NWS-type procedure that is less rigorous but considerable faster than the full-blown version.

Using this program, a query sequence a couple of hundred residues long can be compared with the whole NBRF protein sequence data bank (some 3500 sequences) in a few minutes on a VAX 11/750 computer, and the search can even be performed in a reasonable time on an IBM PC. A typical output is shown in Fig. 6.2. The statistical significance of a match can be roughly estimated by calculating a z value:

$$z = (\text{similarity score} - \text{mean score from data base scan}) / (\text{standard deviation from data base scan}) \quad (6.4)$$

Lipman and Pearson offer the following guidelines: $z > 3$, possibly significant; $z > 6$, probably significant; $z > 10$, significant.

A protein sequence data bank search is both faster and more sensitive than a DNA bank search, and should be chosen if possible. Also, since the computing time drops quickly with increasing *k*, a FASTP scan would first use $k = 2$, and only if no clearly related sequences appear would one make a run with $k = 1$ (which will consume about five times as much CPU time).

IV. MULTIPLE ALIGNMENTS

As remarked already by Needleman and Wunsch (1970), the NWS algorithm can be generalized to allow simultaneous alignment of more than two sequences. However, unless special measures are adopted, this will require impossibly long computations. Murata *et al.* (1985) present one solution to this problem where three sequences are optimally aligned in the NWS sense. The computing time grows as the cube of the mean sequence length, rather than as the fifth power as in the original method. A drawback of the procedure is that the gap weight is independent of the gap length (i.e., $W(k) =$ const.), but this restriction can be removed, as shown by Gotoh (1986).

If one is satisfied with a similarity score rather than a full alignment, one can look for subsequence matches and compare the length distribution of these matches with the random expectation. A program written by Karlin *et al.* (1983) finds all direct repeats in a sequence that can be many tens of kilobases long; similarities between different sequences can be found by concatenating the sequences and searching for repeats in the extended sequence. This program is based on a word-search algorithm, and can also find dyad symmetries.

More heuristic algorithms that will align many sequences simultaneously have also been proposed. One idea is to construct a consensus by first comparing two sequences and extracting a first consensus, then compare this consensus to a third sequence, etc. Bacon and Anderson (1986) thus store the 1000 best subsequence alignments of fixed length (typically 20 residues, gaps are not allowed) from the first comparison, these are then checked against the third sequence, and the 1000 best alignments are stored, and so on. Since only a single pass through the sequences to be compared is made, the method is not fully self-consistent, and the order in which the comparisons are made can influence the results. Nevertheless, for groups of no more than four or five sequences one seems to get fairly consistent similarity scores and close to optimal subsequence alignments.

Obviously, a scheme like this can be developed into an iterative procedure, where each sequence in turn is compared to the emerging

```
bovine opsin, 348 amino acids vs PROTSEQ library

< 2      1 :=
  4      0 :
  6      1 :=
  8      7 :====
 10     30 :===============
.12     54 :===========================
 14    109 :==================================================
 16    260 :==================================================
 18    387 :==================================================
 20    504 :==================================================
 22    535 :==================================================
 24    517 :==================================================
 26    343 :==================================================
 28    302 :==================================================
 30    273 :==================================================
 32    154 :==================================================
 34    107 :==================================================
 36     75 :======================================
 38     43 :======================
 40     48 :========================
 42     11 :======
 44     13 :=======
 46      7 :====
 48      7 :====
 50      4 :==
 52      0 :
 54      1 :=
 56      1 :=
 58      1 :=
 60      0 :
 62      0 :
 64      0 :
 66      0 :
 68      0 :
 70      0 :
 72      0 :
 74      0 :
 76      0 :
 78      0 :
 80      0 :
> 80     5 :===
 891051 residues in  3800 sequences, mean score:  23.1 (6.09)
  750 scores better than 28 saved, ktup: 2, fact: 8  scan time:  0:02:55.58
```

Fig. 6.2. Typical FASTP output. The histogram shows the distribution of initial scores for the whole data bank.

```
The best scores are:                                                 init,  opt
OOBO     :  Rhodopsin - Bovine                                       1865, 1865
OOHUB    :  Blue-sensitive opsin - Human                              861,  935
OOSH     :  Rhodopsin - Sheep (fragments)                             756, 1017
OOHUR    :  Red-sensitive opsin - Human                               706,  799
OOHUG    :  Green-sensitive opsin - Human                             704,  800
QQBE50   :  Hypothetical BNLF1 protein - Epstein-Barr vir              57,   67
AHRB     :  Ig alpha chain C region - Rabbit (fragment)                56,   56
SAVLVD   :  Probable major surface antigen precursor - He              53,   64
DELOG3   :  Glyceraldehyde 3-phosphate dehydrogenase (EC               50,   50
JDVLD    :  Probable DNA polymerase (EC 2.7.7.7) - Duck h              50,   79
WGECH    :  Hygromycin B phosphotransferase (EC 2.7.1.-)               49,   49


OOHUB    :  Blue-sensitive opsin - Human                              861,  935
 44.4% identity in 349 aa overlap

                                          :
    1' MNGTEGPNFYVPFSNKTGVVRSPFEAPQYYLAEPWQFSMLAAYMFLLIMLGFPINFLTLY
       :.  .. .::. :.: ..:   :...:::..:  :.: . ::.:  ....:::.: ..:
    1" MRKMSEEEFYL-FKNISSVG--PWDGPQYHIAPVWAFYLQAAFMGTVFLIGFPLNAMVLV
                                          :

   61' VTVQHKKLRTPLNYILLNLAVADLFM-VFGGFTTTLYTSLHGYFVFGPTGCNLEGFFATL
       .:...:::: ::::::.:.. ..... .:. : ... .: .::::::.  :.:::::.:.
   58" ATLRYKKLRQPLNYILVNVSFGGFLLCIFSVF-PVFVASCNGYFVFGRHVCALEGFLGTV

  120' GGEIALWSLVVLAIERYVVVCKPMSNFRFGENHAIMGVAFTWVMALACAAPPLVGWSRYI
       .: .. :::. ::.:::.:::::..:::::...::.   :   ::...... . ::. ::::.:
  117" AGLVTGWSLAFLAFERYIVICKPFGNFRFSSKHALTVVLATWTIGIGVSIPPFFGWSRFI

  180' PEGMQCSCGIDYYTPHEETNNESFVIYMFVVHFIIPLIVIFFCYGQLVFTVKEAAAQQQE
       :::.::::: :.::  .. ..::.. ..:.  ::.:: .: :.:.::. ..:..::::::
  177" PEGLQCSCGPDWYTVGTKYRSESYTWFLFIFCFIVPLSLICFSYTQLLRALKAVAAQQQE

  240' SATTQKÄEKEVTRMVIIMVIAFLICWLPYAGVAFYIFTHQGSDFGPIFMTIPAFFAKTSA
       ::::::::.::.:::::..:: .: .:..:::. :.:. .... ...  ..:::.::.:..
  237" SATTQKAEREVSRMVVVMVGSFCVCYVPYAAFAMYMVNNRNHGLDLRLVTIPSFFSKSAC

                          :
  300' VYNPVIYIMMNKQFRNCMVTTLCCGKNPLGDDEASTTVSKTETSQVAPA .
       .::::.:: .::::::..:.. .. ::: ...:.... .  :::.: :...
  297" IYNPIIYCFMNKQFQACIM-KMVCGK-AMTDESDTCSSQKTEVSTVSSTQVGPN
                          :

Library scan:  0:02:55.58  total CPU time:  0:03:39.84
```

Fig. 6.2 (Continued)

consensus sequence until the consensus "converges" to a stable sequence that remains unchanged when compared to all sequences in the set. This idea has been implemented in the MULTAN program (Bains, 1986), which can align large sets of DNA sequences (up to 50 sequences of ~1 kb each) in the following manner. First, a trial consensus is chosen among the sequences to be aligned. Second, pairwise alignments between the trial consensus and every sequence in the set are constructed (this could in principle be done by a NWS procedure, but MULTAN works with a faster and less rigorous algorithm). Third, gaps that have been introduced into the trial consensus in the pairwise alignments are likewise introduced into all sequences in the set. Fourth, the sequences in the set (now of equal lengths and including gaps) are written one above the other, and a new trial consensus is constructed from the incidence of each kind of base in each position in the aligned set. This process is repeated until a stable consensus emerges.

Apparently, the final alignments do not depend on the order in which the sequences are compared, or which sequence is chosen as the first trial consensus. For large sets of fairly diverged or small sets of very diverged sequences, no stable consensus may be reached; there are a number of parameters that affect the alignment subroutine (step two) that can be varied to overcome this problem, but only at the cost of loosing some fine structure in the final alignment (basically, many positions in the consensus will be of the type "no nucleotide preferred").

V. MEASURES OF AMINO ACID SIMILARITY

In the examples considered above, we have only scored identities and mismatches, i.e., $\beta(a_i,b_j) = 0, 1$. This choice is natural when one is aligning DNA sequences (possibly, though, in some situations transitions and transversions could be differently weighted), but when one is dealing with protein sequences it will often be appropriate to distinguish between conservative and nonconservative replacements. A number of similarity matrices $\beta(i,j)$ assigning weights to all possible amino acid substitutions have been proposed over the years; a few of the most commonly used will be reviewed in this section.

A simple qualitative grouping of the 20 naturally occurring amino acids is sometimes used when one is mainly interested in the overall chemical or structural similarity of two sequences. Various classifications are possible; e.g., Karlin and Ghandour (1985), who suggest the following four "alphabets":

Chemical: acidic (Asp, Glu); aliphatic (Ala, Gly, Ile, Leu, Val); amide (Asn, Gln); aromatic (Phe, Trp, Tyr); basic (Arg, His, Lys); hydroxyl (Ser, Thr); imino (Pro); sulfur (Cys, Met)

Functional: acidic, basic, hydrophobic (Ala, Ile, Leu, Met, Phe, Pro, Trp, Val); polar (Asn, Cys, Gln, Gly, Ser, Thr, Tyr)

Charge: acidic, basic, neutral

Structural: ambivalent (Ala, Cys, Gly, Pro, Ser, Thr, Trp, Tyr); external (Arg, Asn, Asp, Gln, Glu, His, Lys); internal (Ile, Leu, Met, Phe, Val).

Groupings such as these can be used first to write the sequences to be compared in one of the alphabets, and then to align them scoring only for identities and mismatches in the chosen alphabet.

A more complete set of groupings along the same lines has been provided by Taylor (1986b), who uses the well-known Venn diagrams from set theory to display various possible groupings and their intersections. Rather than using his groupings to align sequences, he applies them *after* the alignment as a way to express graphically the varying degree of residue conservation along the sequences. This is achieved by plotting the number of primitive groups needed to encompass all residues found in a given position as a function of the position in the chain: Highly conserved positions will be represented by one or two groups, whereas divergent positions will need many groups to encompass all residues found.

Most of the more commonly used alignment programs use similarity matrices with graded scales, however. Feng *et al.* (1985) have conducted a thorough comparison of four widely used matrices: the Unitary Matrix (UM), the Genetic Code Matrix (GC), the Structure–Genetic Matrix (SG), and Dayhoff's Log-Odds Matrix (LO). UM scores only identities and mismatches, i.e., the scoring matrix is diagonal. GC scores are derived from the minimum number of base changes needed to convert a residue into another (zero, one, two, and three base changes are given the weights 4, 2, 1, 0; note that maximum similarity rather than minimum distance is measured here). The SG and LO scores, finally, are based on observed frequencies of amino acid replacements in evolutionary related proteins (see Table 6.2).

From their study, Feng *et al.* conclude that "on the average, weight-

TABLE 6.2. **Similarity Values According to the Structure-Genetic (Upper Triangle) and Log-Odds Matrices (Lower Triangle)[a]**

Residue	C	S	T	P	A	G	N	D	E	Q	H	R	K	M	I	L	V	F	Y	W	
	6	4	2	2	2	3	2	1	0	1	2	2	0	2	2	2	2	3	3	3	C
		6	5	4	5	5	5	3	3	3	3	3	3	1	2	2	2	3	3	2	S
C	20		6	4	5	2	4	2	3	3	2	3	4	3	3	2	3	1	2	1	T
S	8	10		6	5	3	2	2	3	3	3	3	2	2	2	3	3	2	2	2	P
T	6	9	11		6	5	3	4	4	3	2	2	3	2	2	2	5	2	2	2	A
P	5	9	8	14		6	3	4	4	2	1	3	2	1	2	2	4	1	2	3	G
A	6	9	9	9	10		6	5	3	3	4	2	4	1	2	1	2	1	3	0	N
G	5	9	8	7	9	13		6	5	4	3	2	3	0	1	1	3	1	2	0	D
N	4	9	8	7	8	8	10		6	4	2	2	4	1	1	1	4	0	1	1	E
D	3	8	8	7	8	9	10	12		6	4	3	4	2	1	2	2	1	2	1	Q
E	3	8	8	7	8	8	9	11	12		6	4	3	1	1	3	1	2	3	1	H
Q	3	7	7	8	8	7	9	10	10	12		6	5	2	2	2	2	1	1	2	R
H	5	7	7	8	7	6	10	9	9	11	14		6	2	2	2	3	0	1	1	K
R	4	8	7	8	6	5	8	7	7	9	10	14		6	4	5	4	2	2	3	M
K	3	8	8	7	7	6	9	8	8	9	8	11	13		6	5	5	4	3	2	I
M	3	6	7	6	7	5	6	5	6	7	6	8	8	14		6	5	4	3	4	L
I	6	7	8	6	7	5	6	6	6	6	6	6	6	10	13		6	4	3	3	V
L	2	5	6	5	6	4	5	4	5	6	6	5	5	12	10	14		6	5	3	F
V	6	7	8	7	8	7	6	6	6	6	6	6	6	10	12	10	12		6	3	Y
F	4	5	5	3	4	3	4	2	3	3	6	4	3	8	9	10	7	17		6	W
Y	8	5	5	3	5	3	6	4	4	4	8	4	4	6	7	7	6	15	18		
W	0	6	3	2	2	1	4	1	1	3	5	10	5	4	3	6	2	8	8	25	

[a]From Feng *et al.* (1985).

ing (beyond the simple UM, my note) can help in establishing distant relationships," and, somewhat hesitatingly, they judge the LO matrix to be the most effective. For sequences with high degrees of similarity (> 30%), all weighting schemes result in almost identical NWS alignments, and simple UM scoring is sufficient.

Other measures of amino acid similarity that have been used to compare previously aligned protein sequences include secondary structure propensities (Pongor and Szalay, 1985), physicochemical characteristics (Bacon and Anderson, 1986), and hydrophobicity values (Sweet and Eisenberg, 1983). The latter derived an Optimal Matching Hydrophobicity (OMH) scale (Table 5.4 in Chapter 5), by taking the hydrophobicity of each kind of residue to be the average hydrophobicity of all residues found to replace it in an amino acid replacement table constructed from evolutionary related sequences by Dayhoff. The averaging procedure was repeated until the scale was self-consistent, i.e., until the values did not change upon further averaging, and it was found that the final scale was largely independent of the details of the starting scale. With scales such as these, one can get measure of the "structural relatedness" of two aligned proteins by calculating a correlation coefficient between the OMH values of all residue pairs, or by visually comparing the hydrophobocity or secondary structure prediction plots for the two proteins.

VI. ESTIMATING THE STATISTICAL SIGNIFICANCE OF AN ALIGNMENT

Estimating the significance of an observed similarity score is still somewhat of a black art—no satisfactory statistical theory that takes care both of the properties of the different alignment algorithms and the statistical properties of unrelated nucleotide sequences (e.g., constraints on dinucleotide frequencies, codon usage, large-scale variations in base frequencies) exist. Techniques commonly used to assess the quality of an alignment include Monte Carlo methods, statistical analyses of large reference samples of presumably unrelated sequences, and approximate analytical formulas derived for particular alignment algorithms under various assumptions of what a "random" sequence is like. Probably because they are easier to deal with theoretically, subsequence alignments seem to have been studied more intensively than global alignments.

A. Subsequence Alignments

Goad and Kanehisa (1982) and Kanehisa (1984) have derived analytical formulae for assessing the significance of similarities found by their DNA subsequence matching algorithm described in Section III,A. They assume that the appearance of the four bases in unrelated sequences are random, and, recognizing this limitation, recommend their method only as a first screening procedure; if the estimated significance of an observed similarity is weak or in any way dubious, more realistic methods should be applied.

A simpler kind of alignment algorithm, used by, e.g., McLachlan (1971) for protein sequence alignment, that does not allow for gaps but simply finds the best subsequence matches of fixed length by sliding the two sequences along one another, is easier to analyze theoretically. McLachlan has derived a double matching probability to describe the statistics of comparing two unrelated sequences in this way, and an analogous n-fold matching probability was recently provided by Bacon and Anderson (1986) as a part of their multiple sequence alignment algorithm (Section IV).

The importance of taking the known statistical properties of natural DNA into account has been nicely demonstrated by Lipman *et al.* (1984). Applying the local similarity algorithm of Smith and Waterman (SW) described in Section III,A above to a large set of randomly chosen sequences from GenBank, as well as to three sets of differently randomized sequences, they could show that an alignment that looks significant when compared with controls that have been randomized by a procedure only preserving the overall base composition does not appear significant when compared with controls that have been randomized by procedures preserving nearest-neighbor frequencies or local base composition.

In a more empirical vein, T. F. Smith *et al.* (1985), again using the SW local alignment algorithm, have calculated the distribution of the maximum similarity scores resulting from all pairwise comparisons of 204 vertebrate DNA sequences from GenBank with each other. From probability theory, they expected that the mean scores for unrelated sequences of lengths n and m would grow as $\log(nm)/\log(1/p)$ where p is the probability of a base match ($p =$ the sum of the squares of the base frequencies); indeed, this turned out to be the case. From a least-square fitting to the observed distribution, they derived the following expressions for the expected mean score S and standard deviation σ:

$$S = 2.55 \times \log(nm)/\log(1/p) - 8.99$$
$$\sigma = 1.78 \tag{6.5}$$

The significance of a similarity score obtained from the SW algorithm with the parameters as used by T. F. Smith *et al.* (match = 1, mismatch = −0.9, gap = −2 per base) can thus be estimated by subtracting S and dividing by σ; this will give the number of standard deviations above the mean of the observed score. Other parameter values and other sets of reference sequences may give slightly different numerical factors in Eq. (6.5), but the basic approach is clearly applicable in all cases.

B. Global Alignments

Concerning global alignments, I will only relay two rules of thumb. The first one is the z-value rules suggested by Lipman and Pearson (1985) for estimating the significance of protein similarity scores obtained by the FASTP program; these have already been given in Section III,B. The second rule applies to protein sequence alignments by the NWS algorithm: When using, e.g., the Unitary Matrix scoring for matches and mismatches and a gap penalty $W(k) = -2.5$ regardless of the size of the gap, Feng *et al.* (1985) find that a normalized alignment score (100 × similarity score/length of shorter sequence) larger than 175 "is a good indication of a genuine relationship." Similar thresholds can be defined for the other scoring systems discussed by these authors (Section V).

Before leaving the subject, let me just note that one does not have to know the sequence of a particular region of the genome to make similarity assessments; restriction maps often contain valuable similarity/homology information. Thus, Waterman *et al.* (1984) have developed an NWS-type algorithm that finds the minimum weighted sum of genetic events (appearance/disappearance of sites; changes in fragment lengths) required to convert one map into the other.

Chapter 7

Machine Molecular Biology: Running Experiments on the Computer

Computers have one big advantage: They are fast and cheap. In molecular biology, this means that one can not only analyze natural sequences, one can also generate hypothetical sequences according to some set of rules and analyze them using the same methods, creating, in effect, very large reference samples that can be used to assess the significance of some perceived sequence pattern. One can also simulate various evolutionary scenarios or aspects of cellular processes such as protein synthesis, and in this way gain a more complete understanding of the underlying mechanisms and their consequences.

In this chapter, I will present some studies of this kind, where computer modeling has been central to the conclusions reached. Following the old rule that the most important lesson should be up front, the first few examples will illustrate that great care should be exercised when making statements about "significant" patterns in a set of sequences. The remainder will briefly touch on some activities that we in the CMB (Computer Molecular Biology) field sometimes indulge in — making ribosomes hop around inside the computer, or turning the memory banks into evolving genomes. Even theoreticians must have some fun.

I. PATTERNS THAT WERE NOT

Spurious similarities (homologies) between DNA or protein sequences are surprisingly common, and it is easy to fool oneself that the base-pairing scheme or sequence alignment one has come up with is indicative of selection at work. The examples given below range from DNA recombination to the distribution of genes in the *Escherichia coli* genome, from amphiphilic transmembrane helices to nucleotide-binding proteins. They are all chosen to make the point that the computer can be an important tool for evaluating the statistical significance of a seemingly convincing observation.

A. Patchy Homologies

Patchy homologies between otherwise unrelated DNAs have been invoked more than once to explain why a specific recombinatorial event took place at the precise location that it did. A patchy homology is composed of a number of short, perfectly homologous segments separated by short, nonhomologous loops, much in the way a folded RNA structure looks. Are such homologies, as they are found in natural DNAs, really significant, or should we expect base-paired structures of comparable extent and stabilities even in random sequences?

Savageau *et al.* (1983) approached this question by analyzing patchy homologies suggested in the literature, using the Zuker and Stiegler RNA-folding algorithm (Chapter 4, Section VII,B). They could show that even more stable structures than the ones originally suggested were often found by the algorithm; nevertheless, random sequences with the same GC content as the natural ones could form equally stable and extensive structures. The concept of patchy homology thus on closer scrutiny turned out not to have much explanatory power.

B. U2 Small Nuclear RNA and Splicing

In their analysis of splice junctions, Nakata *et al.* (1985) made a similar observation regarding a proposed base-pairing between U2 snRNA and the exon sequences flanking the splice site (Chapter 4, Section IV,B). By comparing a set of true and false splice sites, they could show that U2 snRNA could make equally stable base pairs to both sets, whereas the U1 snRNA, which was known to pair with the 5′ end of introns, in fact appeared to match the true splice sites better (Fig. 7.1).

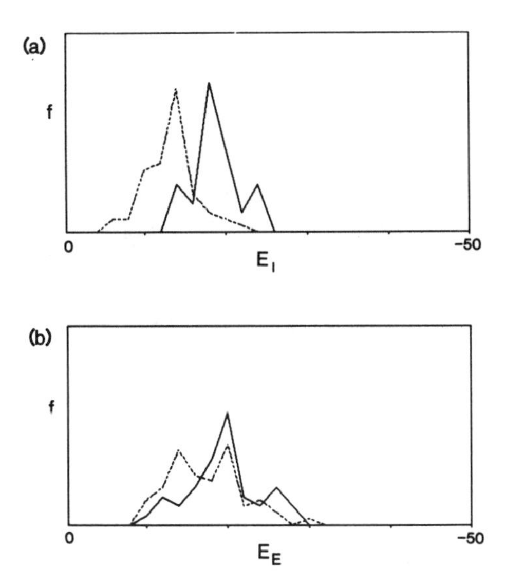

Fig. 7.1. Distribution of optimal base-pairing energies calculated for the pairing of U1 snRNA (a) and U2 snRNA (b) with a set of "true" (solid line) and "false" (dashed line) splice sites. (From Nakata *et al.*, 1985.)

C. Gene Distribution in *E. coli*

A more complicated case concerns the gene density and a perceived spatial symmetry in the gene distribution over the *E. coli* chromosome (Jurka and Savageau, 1985). Looking at a circular map of the genome, one can discern a symmetry axis that divides the map into two halves with more or less mirror-symmetrical distributions of regions of high and low gene density, suggesting that the chromosome may have resulted from one or more duplications of a primordial, much smaller genome. However, by first showing that the gene density distribution (number of loci per 0.5 min) was well reproduced by a skewed, so-called lognormal distribution, and then generating random maps respecting this density distribution, Jurka and Savageau demonstrated that equally convincing symmetries were present also in the random maps. In fact, an axis of quasi-symmetry is not very unlikely to exist even in the general case, since the great freedom allowed in placing the axis introduces a strong bias in favor of the symmetry one set out to find.

D. Amphiphilic Helices

The same problem is encountered in a very different context, that of amphiphilic helices in proteins (Chapter 5, Section V,C). Recall that an amphiphilic helix is an α-helix with one apolar and one polar and/or charged face. Such helices are often found in membrane-spanning proteins, or in peptides with surface-binding properties (peptide hormones, lytic peptides). The helical wheel representation (Fig. 7.2), is well suited to show the typical clustering of polar/apolar residues on opposite sides of the helix, and has often been used to identify putative amphiphilic helices. As in the previous case, however, the great freedom allowed in placing the optimal cutting plane makes it rather probable that one can find seemingly convincing amphiphilic helices even in randomly generated sequences. Using Monte Carlo simulation, Flinta *et al.* (1983) could show that in an 18-residues-long helix, with 6 polar residues randomly distributed on the helical wheel, the probability that all 6 polar residues would cluster on one face is as large as 6%. With 7 polar residues, there is a 2% probability that all 7 will be on one face, and a 22% probability that 6 out of 7 are clustered on one face. The graphical cutting plane method is thus not very reliable, and

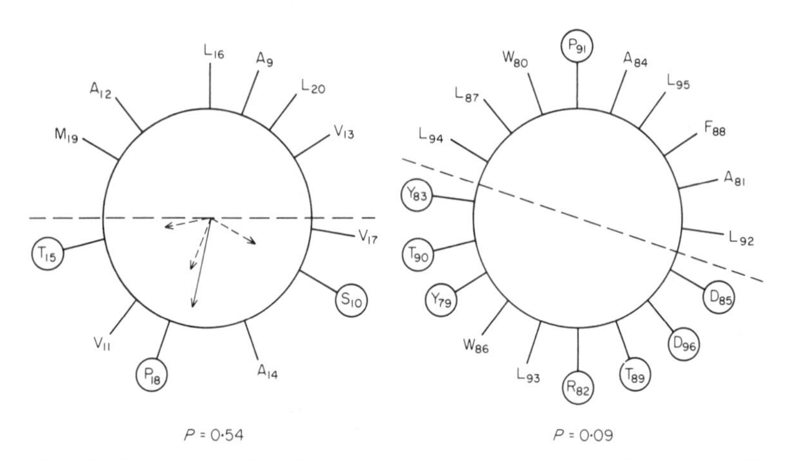

Fig. 7.2. Helical wheel plots of two membrane-spanning segments from phage M13 coat protein (left) and the bacteriorhodopsin C helix (right). Polar and charged residues are encircled. The probability that 3 out of 3 polar residues should cluster on one side in a randomly generated 12-residue helix is 0.54 (cf. the M13 helix); the probability that 7 or 8 out of 8 polar residues should cluster on one side in a random 18-residue helix is 0.09 (cf. the bacteriorhodopsin helix). (From Flinta *et al.*, 1983.)

identification of amphiphilic helices should be based on the hydrophobic moment or some similar measure (Chapter 5, Section V,C). To assess the significance of the putative helix, one can also generate randomly shuffled versions of the sequence, and calculate the expected distribution of the hydrophobic moment for the randomized sample (von Heijne, 1986d).

E. Weak Homologies

In general, weak or highly degenerate patterns of sequence similarity should be viewed with suspicion. If such a pattern is thought to be characteristic of a group of functionally, structurally, or evolutionary related proteins, a good test is to search a major data bank such as the NBRF Protein Sequence Data Bank for instances of the observed pattern. A too unrestrictive pattern will fill the screen with matches from a host of unrelated proteins.

Argos and Leberman (1985) and von Heijne and Uhlén (1987) provide examples of this. The first study deals with a suggested pattern common to a number of oncogene products and known nucleotide binding domains: Gly-X-Gly-X-X-Gly-X_{11-17}-Hf-Hf-X-Lys, where Hf is a hydrophobic residue and X is any residue. When the NBRF data bank was searched with this and similar patterns, many unrelated sequence were found, and the authors concluded that "caution should be exercised in drawing conclusions based on homology data obtained by comparing spans of only 15–35 amino acids in length."

Von Heijne and Uhlén (1987) did a similar study of a previously proposed octapeptide cell-surface binding domain from *Staphylococcus aureus* protein A supposed to be diagnostic of membrane-interacting proteins in general; again, the result of the data bank search revealed a large number of matches from proteins that have nothing to do with cell surfaces or membranes.

II. COMPUTER-GUIDED INTUITION

Many postulated processes and mechanism in molecular biology can be "tested" and explored through computer simulation. Here, I will only describe a couple of randomly chosen examples to illustrate the approach and the semiquantitative character of the results one might get. It is not only that one can explore complicated kinetic schemes and ranges of parameter variation in this way; a perhaps equally important point is that the conversion of a woolly mechanistic hypothesis into a precise computer algorithm forces one to confront ambiguities and unstated assumptions in the original model.

A. Genome Evolution

In a rather neat illustration of evolution at work, Loomis and Gilpin (1986) describe a computer simulation of a genome evolving by random duplication and deletion events. Vital genes are postulated to be made up of four units: a promoter (#), two coding units (AA), and a terminator (.), thus written #AA.. Duplication of the whole gene results in two functional copies; duplication of less than the whole gene results in nonfunctional sequences (dashes), as does a deletion of a part of a gene. A deletion in a single-copy gene kills the genome. Genes present in multiple copies can mutate into new and different genes: #AA.→#AB., etc.

Running this model on the computer, one can follow the evolution of a single genome, and one can collect statistics to describe the process. Under the assumptions of the model, large genomes with a large proportion of vestigial or junk DNA, and with gene families of widely different copy numbers, are a natural outcome (Fig. 7.3). The size of the genome stabilizes when a sufficient proportion of the genome is dispensable junk; only then will most deletions not affect vital genes, and hence allow deletions to balance duplications.

B. Protein Synthesis

If this genome evolution model is perhaps most aptly described as being of the "parlor game" variety—fun but fictitious—a little more can be said for attempts to investigate the kinetics of protein synthesis and ribosome movement with the aid of computer simulation. The recourse to computer modeling in this case was precipitated by the realization that ribosome movement on an mRNA was far too complex to be described analytically. Hence various simulations have been performed over the past 15 years (reviewed in von Heijne *et al.*, 1987), all basically having one or more mRNAs in the computer's memory over which ribosomes move stochastically according to a random number generator. Since the positions of the individual ribosomes are kept track of, it is easy to include rules that prevent ribosomes from

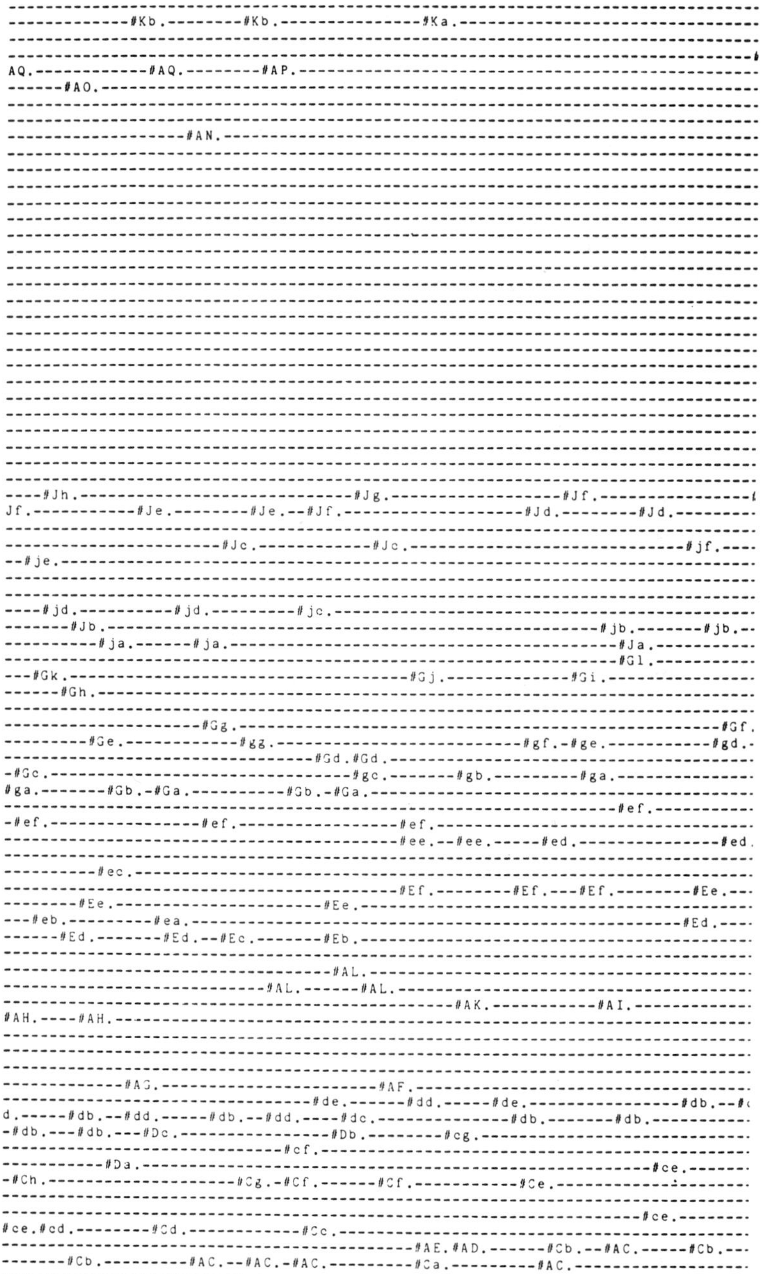

```
--------------------------------------------------------------------------------
-------------#Kb.--------#Kb.---------------#Ka.--------------------------------
--------------------------------------------------------------------------------
-------------------------------------------------------------------------------#
AQ.------------#AQ.--------#AP.-------------------------------------------------
------#AO.----------------------------------------------------------------------
--------------------------------------------------------------------------------
--------------------------------------------------------------------------------
-------------------#AN.---------------------------------------------------------
--------------------------------------------------------------------------------
--------------------------------------------------------------------------------
--------------------------------------------------------------------------------
--------------------------------------------------------------------------------
--------------------------------------------------------------------------------
--------------------------------------------------------------------------------
--------------------------------------------------------------------------------
--------------------------------------------------------------------------------
--------------------------------------------------------------------------------
--------------------------------------------------------------------------------
--------------------------------------------------------------------------------
--------------------------------------------------------------------------------
--------------------------------------------------------------------------------
--------------------------------------------------------------------------------
--------------------------------------------------------------------------------
--------------------------------------------------------------------------------
--------------------------------------------------------------------------------
--------------------------------------------------------------------------------
--------------------------------------------------------------------------------
--------------------------------------------------------------------------------
--------------------------------------------------------------------------------
----#Jh.-----------------------------#Jg.------------------#Jf.----------------#
Jf.-----------#Je.--------#Je.--#Jf.-------------------#Jd.--------#Jd.---------
--------------------------------------------------------------------------------
-----------------------#Jc.------------#Jc.-----------------------------#jf.----
--#je.--------------------------------------------------------------------------
--------------------------------------------------------------------------------
--------------------------------------------------------------------------------
----#jd.----------#jd.---------#jc.---------------------------------------------
-------#Jb.----------------------------------------------------#jb.-------#jb.--
----------#ja.------#ja.-----------------------------------------#Ja.-----------
-----------------------------------------------------------------#Gl.-----------
---#Gk.------------------------------------#Gj.-------------#Gi.----------------
------#Gh.----------------------------------------------------------------------
--------------------------------------------------------------------------------
---------------------#Gg.---------------------------------------------------#Gf.
---------#Ge.------------#gg.--------------------------#gf.-#ge.-----------#gd.-
---------------------------------#Gd.#Gd.---------------------------------------
-#Gc.--------------------------------#gc.-------#gb.---------#ga.---------------
#ga.-------#Gb.-#Ga.----------#Gb.-#Ga.-----------------------------------------
-----------------------------------------------------------------#ef.-----------
-#ef.----------------#ef.-----------------#ef.----------------------------------
------------------------------------------#ee.--#ee.-----#ed.---------------#ed.
--------------------------------------------------------------------------------
----------#ec.------------------------------------------------------------------
------------------------------------------#Ef.--------#Ef.---#Ef.--------#Ee.---
--------#Ee.----------------------#Ee.------------------------------------------
---#eb.---------#ea.----------------------------------------------------#Ed.----
------#Ed.-------#Ed.--#Ec.-------#Eb.------------------------------------------
--------------------------------------------------------------------------------
-----------------------------------#AL.-----------------------------------------
----------------------------#AL.------#AL.--------------------------------------
------------------------------------------------#AK.-----------#AI.-------------
#AH.----#AH.--------------------------------------------------------------------
--------------------------------------------------------------------------------
--------------------------------------------------------------------------------
--------------------------------------------------------------------------------
-------------#AG.-----------------------#AF.------------------------------------
---------------------------------#de.------#dd.-----#de.----------------#db.--#
d.-----#db.--#dd.-----#db.--#dd.----#dc.--------------#db.-------#db.-----------
-#db.---#db.---#Dc.----------------#Db.--------#cg.-----------------------------
------------------------------#cf.----------------------------------------------
-----------#Da.------------------------------------------------------#ce.-------
-#Ch.--------------------#Cg.-#Cf.------#Cf.-----------#Ce.---------------------
--------------------------------------------------------------------------------
--------------------------------------------------------------------#ce.--------
#ce.#cd.--------#Cd.------------#Cc.--------------------------------------------
--------------------------------------------#AE.#AD.------#Cb.--#AC.-----#Cb.---
-------#Cb.---------#AC.--#AC.-#AC.---------#Ca.---------#AC.-------------------
--------------------------------------------------------------------------------
-------------------------#AA.-------
```

Fig. 7.3. Complex computer-generated genome after 10,000 accumulated events. Dashes are vestigial sequences (junk DNA). (From Loomis and Gilpin, 1986.)

colliding, or that check whether a given self-complementary stretch of mRNA is prevented or not from folding into a hairpin loop by the ribosomes. The effects of ribosome density, RNA structure, and codon usage on the overall kinetics of protein synthesis have been investigated in this way, and the relations between experimentally accessible quantities such as overall rate of synthesis or polysome size distribution, and intrinsic parameters (rate constants for initiation, elongation, and termination) have been made more precise.

C. Attenuation Control

In bacteria, a number of amino acid biosynthetic operons are controlled by an attenuator mechanism that directly determines whether or not RNA polymerase transcribes through an early potential termination site. A ribosome translating a short leader peptide region in the early parts of the transcript is thought either to stall at a critical codon corresponding to a rare aminoacyl-tRNA, thus interfering with the formation of a particular hairpin loop in the mRNA and preventing transcription termination, or else to translate the whole leader peptide and allow termination (Fig. 7.4) (Yanofsky, 1981; Carter *et al.,* 1986).

This control mechanism integrates a number of the themes discussed in this chapter: overall kinetics of protein synthesis, modulation of the elongation rate through aminoacyl-tRNA availability, and ribosomes influencing mRNA folding. It therefore comes as no great surprise that it has been the subject of theoretical modeling. Manabe (1981) was the first to derive an expression for the probability that a ribosome will be stalled at one of the critical codons when the polymerase reaches the termination site, and construced theoretical response curves for the attenuator under various degrees of amino acid starva-

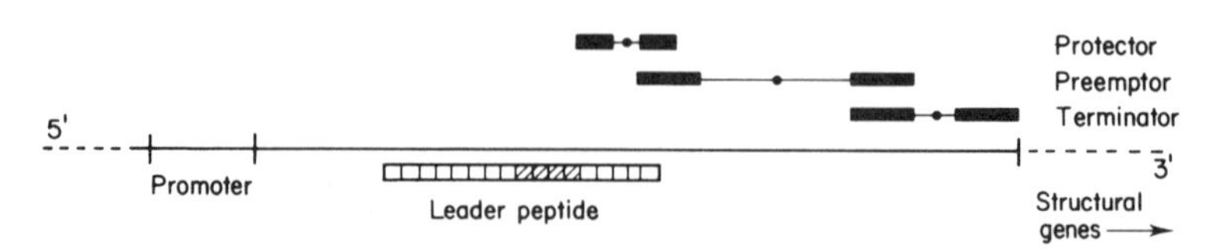

Fig. 7.4. Regulatory region of an attenuator-controlled operon. The critical control codons in the leader peptide are hatched, and self-complementary regions are shown in black. A ribosome stalled on one of the control codons in the leader transcript will prevent protector, but not preemptor, formation, and hence make terminator formation impossible. (From von Heijne, 1982.)

tion. This was extended by von Heijne (1982), who simulated the effects of mutations in the critical codons and discussed the design principles of the attenuator region based on the results from the computer calculations.

More recently, Suzuki *et al.* (1986) have incorporated an experimentally detected transcriptional pausing of the RNA polymerase at a terminator-like region in the attenuator into the theoretical model, and have shown that this pause, which leads to better synchronization between ribosome and polymerase movement (Fisher *et al.,* 1985), makes the control mechanism much more responsive to changes in aminoacyl-tRNA concentrations and less sensitive to stochastic variations in the timing of ribosome-binding to the transcript.

With these examples, picked on the basis of personal familiarity rather than far-reaching relevance, I hope to have made a case for sometimes venturing beyond simple sequence analysis into more complex kinds of modeling in molecular biology. Although most often relegated to the pages of the *Journal of Theoretical Biology,* works of this kind can hold important messages also for the hard-core experimentalist, at least insofar as it shows him or her how well or poorly the molecular mechanisms supposed to underlie a particular process perform when formulated as a precise, quantitative model. Understood in this way, theoretical molecular biology, far from being an unnecessary intrusion into a fundamentally experimental field, can perhaps be seen as a legitimate means of sharpening the ideas that guide us all.

Chapter 8

Sequences: Where Theoretical and Experimental Molecular Biology Meet

In a very real sense, molecular biology is all about sequences. First, it tries to reduce complex biochemical phenomena to interactions between defined sequences—either protein or polynucleotide, sometimes carbohydrates or lipids—then, it tries to provide physical pictures of how these sequences interact in space and time. Theoretical sequence analyses of various kinds are important in this context, since many of the relevant patterns are not immediately obvious. The sheer amount of sequence data now available also makes the appearance of a new breed of "theoretical molecular biologists" unavoidable, and if current proposals to sequence "the" human genome (as if this were a well-defined entity) are put into practice, automated data handling and, above all, data analysis must be given a very high priority. If this book has done nothing else, I hope that it has shown how far we still are from algorithms that will, in the absence of auxiliary information, predict *anything* about new sequences with a reliability even approaching, say, 95% accuracy. If, as by a miracle, 10^9 bp of human DNA would suddenly appear in GenBank, we might be able to find most of the restriction sites, but beyond that it would be a mess (and we would need all the CRAYs in the world just to find that out).

Given this impetus, I think it is a fair bet that computerized molecular biology (CMB) will attract many more workers in the near future. However, if any lesson is to be drawn from what has happened in the somewhat amorphous field of theoretical biology in the past, it surely is that to be able to make a useful contribution one must first and foremost be a biologist, and only second a theoretician. Theoretical biology does not have the internal coherence and unifying theoretical framework of theoretical physics, and cannot expect to entertain the same privileged position *vis à vis* her experimental companion. We in the CMB field must try much harder to communicate with our colleagues in the lab, to speak their language: They are our principal audience.

Sequence analysis provides a good point of contact. Experimentalists know sequences only too well; sequences are easy to handle on a computer, and if CMB can prove its worth by helping to decipher their messages and by providing tools for fast and reliable scanning procedures, we are in business. We have to develop better algorithms, we have to find ways to cope with massive amounts of data, and above all we have to become better biologists. But that's all it takes.

Appendix 1

Molecular Biology Databases

The following list of databases was derived from LiMB (Listing of Molecular Biology databases), a database at Los Alamos National Laboratory providing information about the many data banks now covering information related to molecular biology. For more information about LiMB, see the entry for this effort in the list below.

The list is based on a questionnaire that was sent out between October 1986 and March 1987, and has been compiled by Frances Martinez and Christian Burks (Theoretical Biology and Biophysics Group, Los Alamos National Laboratory).

The first entry in the list indicates what information is given in each field for the following entries.

entry	Name of entry (acronym/initials for database)
gen.nam	Name for general inquiry
gen.add	Address for general inquiry
gen.tel	Telephone number for general inquiry
name.now	Formal name of database
data.pri	Primary data items in the database

entry	AANSPII
gen.nam	Elvin Kabat
gen.add	National Institutes of Health Bldg. 8, Room 126 Bethesda, MD 20892 USA
gen.tel	301-496-0316
name.now	Amino Acid and Nucleotide Sequences of Proteins of Immunological Interest

data.pri	nucleotide and amino acid sequences for immunoglobulins and related proteins
entry	BRD
gen.nam	Volker Erdmann
gen.add	Institut für Biochemie FB Chemie Freie Universität Berlin Otto-Hahn-Bau Thielallee 63 D-1000 Berlin 33 (Dahlem) FEDERAL REPUBLIC OF GERMANY
gen.tel	030-838-60-02
name.now	Berlin RNA Data Bank
data.pri	5 S rRNA nucleotide sequences; 5.8 S rRNA nucleotide sequences; 4.5 S rRNA nucleotide sequences
entry	CSD
gen.nam	Peter Albersheim
gen.add	Complex Carbohydrate Research Center Russell Lab P.O. Box 5677 Athens, GA 30613 USA
gen.tel	404-546-3312
name.now	Carbohydrate Structure Database
data.pri	polysaccharide sequences
entry	CSRS
gen.nam	Ram Reddy
gen.add	Department of Pharmacology Baylor College of Medicine Houston, TX 77030 USA
gen.tel	713-799-4458, ext. 26
name.now	Compilation of Small RNA Sequences
data.pri	nucleotide sequences for small RNAs not directly involved in protein synthesis (e.g., snRNAs and scRNAs)
entry	CUTG
gen.nam	Toshimichi Ikemura
gen.add	National Institute of Genetics Mishima Shizuoka 411 JAPAN
gen.tel	0559-75-0771, ext. 355

name.now	Codon Usage Tabulation from GenBank
data.pri	frequency of codon use in individual genes
entry	DDBJ
gen.nam	Sanzo Miyazawa
gen.add	National Institute of Genetics Mishima 411 JAPAN
gen.tel	0559-75-0771
name.now	DNA Data Bank of Japan
data.pri	nucleotide sequences
entry	DRHPL
gen.nam	Donna R. Maglott
gen.add	American Type Culture Collection 12301 Parklawn Drive Rockville, MD 20852-1776 USA
gen.tel	301-231-5586
name.now	Database for the Repository of Human Probes and Libraries
data.pri	chromosome-specific libraries and human genomic and cDNA clones maintained by ATCC
entry	EMBL
gen.nam	EMBL Data Library
gen.add	Postfach 10.2209 6900 Heidelberg FEDERAL REPUBLIC OF GERMANY
gen.tel	06221-387-258
name.now	The EMBL Data Library
data.pri	nucleotide sequences
entry	GenBank
gen.nam	GenBank
gen.add	BBN Laboratories Inc. 10 Mounton Street Cambridge, MA 02238 USA
gen.tel	617-497-2751
name.now	The GenBank Genetic Sequence Data Bank
data.pri	nucleotide sequences
entry	HDB
gen.nam	Esther J. Asaki

gen.add	Hybridoma Data Bank 12301 Parklawn Drive Rockville, MD 20852 USA
gen.tel	301-231-5586
name.now	Hybridoma Data Bank: A Data Bank on Immunoclones
data.pri	hybridomas; monoclonal antibodies; immunoclones
entry	HGML
gen.nam	Iva H. Cohen
gen.add	Human Gene Mapping Library 25 Science Park New Haven, CT 06511 USA
gen.tel	203-786-5515
name.now	The Howard Hughes Medical Institute Human Gene Mapping Library
data.pri	mapped human genomic loci; DNA probes; RFLPs
entry	LiMB
gen.name	Christian Burks
gen.add	Theoretical Biology and Biophysics Group T-10, MS K710 Los Alamos National Laboratory Los Alamos, NM 87545 USA
gen.tel	505-667-6683
name.now	Listing of Molecular Biology Databases
data.pri	information about molecular biology and related databases
entry	MEDLINE
gen.nam	MEDLARS Management Section
gen.add	National Library of Medicine Bldg. 38A National Institutes of Health Bethesda, MD 20894 USA
gen.tel	800-638-8480
name.now	MEDLINE and Backfiles
data.pri	citations to and Medical Subject Headings (MESH) for the medical/biological literature
entry	NEWAT
gen.nam	R. F. Doolittle
gen.add	Chemistry Department D-006 University of California, San Diego

	La Jolla, CA 92093
	USA
gen.tel	619-534-4417
name.now	NEWAT
data.pri	amino acid sequences
entry	PDB
gen.nam	Frances C. Bernstein
gen.add	Chemistry Department
	Brookhaven National Laboratory
	Upton, NY 11973
	USA
gen.tel	516-282-4384
name.now	Protein Data Bank
data.pri	atomic coordinates for biomolecular structures
entry	PIR
gen.nam	Kathryn E. Sidman
gen.add	National Biomedical Research Foundation
	3900 Reservoir Road, N.W.
	Washington, D.C. 20007
	USA
gen.tel	202-625-2121
name.now	National Biomedical Research Foundation
	Protein Identification Resource
	Protein Sequence Database
data.pri	amino acid sequences
entry	PPR
gen.nam	Esther Lederberg
gen.add	Department of Medical Microbiology
	Sherman Fairchild Building—5402
	Stanford University School of Medicine
	Stanford, CA 94305-2499
	USA
gen.tel	415-723-1772
name.now	Plasmid Prefix Registry
data.pri	plasmid designation; plasmid names
entry	PRF/SEQDB
gen.nam	Yasuhiko Seto
gen.add	Protein Research Foundation
	4-1-2 Ina, Minoh-shi
	Osaka 562
	JAPAN

gen.tel	—
name.now	PRF/SEQDB
data.pri	amino acid sequences
entry	PseqIP
gen.nam	Isabelle Sauvaget
gen.add	Institut Pasteur 25 rue du Docteur Roux 75724 Paris Cedex 15 France
gen.tel	45-68-85-09
name.now	PseqIP
data.pri	amino acid sequences
entry	PTG
gen.nam	James Fickett
gen.add	Theoretical Biology and Biophysics Group T-10, MS K710 Los Alamos National Laboratory Los Alamos, NM 87545 USA
gen.tel	505-665-0479
name.now	Protein Translation of GenBank
data.pri	amino acid sequences
entry	QTDGPD
gen.nam	Cecile Chang
gen.add	Cold Spring Harbor Laboratory P.O. Box 100 Cold Spring Harbor, NY 11724 USA
gen.tel	516-367-8356
name.now	Quest 2D Gel Protein Database
data.pri	spots on two-dimensional protein separation gels
entry	RED
gen.nam	Rich Roberts
gen.add	Cold Spring Harbor Laboratory Cold Spring Habor, NY 11724 USA
gen.tel	516-367-8388
name.now	Restriction Enzyme Database
data.pri	restriction enzymes
entry	SIGPEP
gen.nam	Gunnar von Heijne

gen.add	Department of Theoretical Physics
	Royal Institute of Technology
	S-100 44 Stockholm
	SWEDEN
gen.tel	8-7877172
name.now	Signal Peptide Compilation
data.pri	amino acid sequences of secretory signal peptides

entry	SWISS-PROT
gen.nam	Amos Bairoch
gen.add	Department de Biochimie Medicale
	C.M.U.
	1 Rue Michel Servet
	1211 Geneva 4
	SWITZERLAND
gen.tel	004122-468758
name.now	SWISS-PROT Protein Sequence Data Bank
data.pri	amino acid sequences

entry	SRRSD
gen.nam	George Fox
gen.add	Department of Biochemical Science
	University of Houston SR-1
	4800 Calhoun
	Houston, TX 77004
	USA
gen.tel	713-749-3980; 713-749-2830
name.now	16 S Ribosomal RNA Sequence Database
data.pri	nucleotide sequences of 16 S ribosomal RNAs

entry	SRSRSC
gen.nam	Rupert De Wachter
gen.add	Department Biochemie
	Universiteit Antwerpen (UIA)
	Universiteitsplein 1
	B-2610 Antwerpen
	BELGIUM
gen.tel	32-3-828-25-28, ext. 269
name.now	Small Ribosomal Subunit RNA Sequence Compilation
data.pri	nucleotide sequences of small ribosomal subunit RNAs

entry	TRSD
gen.nam	Mathias Sprinzl
gen.add	Department of Biochemistry
	University of Bayreuth

	Universitätsstrasse 30
	D-8580 Bayreuth
	FEDERAL REPUBLIC OF GERMANY
gen.tel	0921-55-2668
name.now	tRNA Compilation
data.pri	nucleotide sequences for transfer RNAs and transfer RNA genes

Related information can also be obtained from the CODATA Task Group on Coordination of Protein Sequence Data Banks, Professor B. Keil, Institut Pasteur, 28 rue du Docteur Roux, F-75015 Paris, France.

Appendix 2

Some Commercially Available Software Packages for Molecular Biology

For details, get in touch with the supplier. Things change fast in this area, and any particulars concerning the DNA and protein analysis programs true today need to be constantly updated. A compendium listing available software packages has recently been published (Rawlings, 1986).

IntelliGenetics, Inc., 1975 El Camino Real, Mountain View, Ca 94040, USA. Also available for academic users through BIONET, c/o IntelliGenetics.
Data bases: GenBank, EMBL, and NBRF sequence libraries. Robert's (1985) restriction enzyme data base. VectorBank restriction maps of common cloning vectors.
Programs: Recombinant DNA simulation, DNA sequencing manager, sequence editor, similarity search and alignment, restriction map generator, DNA and protein analysis programs, data base search program.
Hardware: IntelliGenetics workstation or via modem

University of Wisconsin Genetics Computer Group, UW Biotechnology Center, 1710 University Avenue, University of Wisconsin, Madison, WI 53705, USA.
Data bases: GenBank, EMBL, and NBRF sequence libraries.
Programs: Sequence editor, DNA and protein analysis, similarity search and alignment, data base search program.
Hardware: VAX/VMS

DNA Star, Inc., 1810 University Avenue, Madison, WI 53711, USA; DNA Star, Inc. Europe, 8 Walpole Gardens, London W4 4HG, England.

Data bases: GenBank, NBRF sequence libraries.
Programs: DNA sequencing manager, sequence editor, restriction map generator, DNA and protein analysis, similarity search and alignment, data base search.
Hardware: IBM PC XT/AT.

PC-Gene, USA: c/o IntelliGenetics (see above); Europe: GenoFit, Case Postale 119, CH-1211 Geneva, Switzerland.
Data bases: EMBL, Swiss-Prot (based on NBRF) sequence libraries.
Programs: Restriction fragment analysis, DNA and protein analysis, similarity search and alignment, data base search.
Hardware: IBM PC XT/AT.

Staden's package, Dr. Roger Staden, Laboratory of Molecular Biology, MRC, University Medical School, Hills Road, Cambridge CB2 2QH, England.
Data bases:—
Programs: DNA sequencing manager, sequence editor, DNA and protein analysis, similarity search and alignment.

GENEUS, Dr. Robert Harr, Unit of Applied Cell and Molecular Biology, Umeå University, S-901 87 Umeå, Sweden.
Data bases: EMBL
Programs: DNA sequencing manager, sequence editor, DNA and protein analysis, similarity search and alignment, data base search.
Hardware: VAX/VMS

The DNA Inspector II, Textco, 27 Gilson Road, West Lebanon, NH 03784, USA.
Data bases:—
Programs: DNA sequencing manager, sequence editor, restriction map generator, DNA and protein analysis, dot matrix.
Hardware: Apple Macintosh

DNASIS, Hitachi America Ltd., 950 Elm Avenue, San Bruno, CA 94066, USA.
Data bases: GenBank, NBRF
Programs: DNA sequencing manager, sequence editor, restriction map generator, DNA and protein analysis, similarity search and alignment, data base search.
Hardware: IBM PC AT/XT

As a concrete example of a typical package (September 1986), the following programs are available from DNA Star, Inc.:
Sequence acquisition
EDITSEQ—sequence editor
READSEQ—input DNA sequence from digitizer
CLONE—splice sequences together

CHCOUNT—count characters in a file
EDITCODE—enter modified genetic codes
SITEFILE—store restriction sites

DNA sequence analysis

MAPSEQ—print sequence, restriction sites, DNA-to-protein translation
LOOPS—find dyad symmetries
MODEL—make a cylindrical projection of B-DNA on paper
BASEDIS—percentages of nucleotides and dinucleotides in scanning window
TRANS—DNA-to-protein translation, codon usage, molecular weight, isoelectric point, charge at given pH
REVTRANS—protein-to-DNA backtranslation, with or without codon bias
FINDCODE—locate coding regions by Fickett's (1982) method
FINDPRO—translate all frames in compact format to facilitate detection of an expected protein segment
ORF—find open reading frames longer than preset limit
QUIKLOOK—rapid printout of sequence

Protein analysis

PROTEIN—predict secondary structure and antigenic sites, plot hydropathy, charge profile, amino acid content, molecular weight
PROBE—select oligonucleotide probes with minimal ambiguity
TITRATE—construct theoretical titration curve

Restriction site analysis

RETABLE—tabulate restriction enzymes and recognition sequences
SITELIST, SITELOOK—search for restriction sites
EDITSITE—add or change restriction enzymes: add recognition sequences such as promoters, etc.
GELWIZE—simulate electrophoretic restriction mapping gels
DIGIGEL—input restriction fragment data from digitizer
SIZEGEL—input restriction fragment data by hand
MAPPER—calculate fit between proposed restriction map and gel data
R-MAP—generate restriction maps automatically from gel data

Sequence comparisons and alignments

COMPARE, AACOMP—find similarities by sliding window method
ALIGN—align two sequences by Wilbur and Lipman (1983) method
SEQCOMP—align by first finding blocks of perfect similarity, then optimizing the alignment by the Needleman–Wunsch (1970) method
GAP—dot-matrix based interactive alignment program

DNA sequencing project management

SEQMAN—keep track of sequencing gels and merge into melds
SEQMANED—editor for the SEQMAN data base
STRATEGY—find optimal strategy for Maxam–Gilbert (1977) sequencing

Data base search

GENEMAN—keyword and sequence pattern search

PROSCAN, NUCSCAN—fast scan of protein and DNA data bases for similarities to a query sequence using the Lipman and Pearson (1985) method

Teletransmission

CALL—telephone modem control program

Bibliography

Aebi, M., Hornig, H., Padgett, R. A., Reiser, J., and Weissmann, C. (1986). Sequence requirements for splicing of higher eukaryotic nuclear pre-mRNA. *Cell* **47**, 555–565.

Almagor, H. (1985). Nucleotide distribution and the recognition of coding regions in DNA sequences: An information theory approach. *J. Theor. Biol.* **117**, 127–136.

Aota, S., and Ikemura, T. (1986). Diversity in G + C content in the third position of codons in vertebrate genes and its cause. *Nucleic Acids Res.* **14**, 6345–6355.

Arentzen, R., and Ripka, W. C. (1984). Introduction of restriction enzyme sites in protein-coding DNA sequences by site-specific mutagenesis not affecting the amino acid sequence: A computer program. *Nucleic Acids Res.* **12**, 777–787.

Argos, P., and Leberman, R. (1985). Homologies and anomalies in primary structural patterns of nucleotide binding proteins. *Eur. J. Biochem.* **152**, 651–656.

Argos, P., and Palau, J. (1982). Amino acid distribution in protein secondary structures. *Int. J. Pept. Protein Res.* **19**, 380–393.

Argos, P., and Rao, J. K. M. (1985). Relationships between exons and the predicted structure of membrane-bound proteins. *Biochim. Biophys. Acta* **827**, 283–297.

Argos, P., Rao, J. K. M., and Hargrave, P. A. (1982). Structural prediction of membrane-bound proteins. *Eur. J. Biochem.* **128**, 565–575.

Armstrong, J., Atenico, E. J., Bergman, B. E., Bilofsky, H. S., Brown, L. B., Burks, C., Cameron, G. N., Cinosky, M. J., Elbe, U., England, C. E., Fickett, J. W., Foley, B. T., Goad, W. B., Hamm, G. H., Hayter, J. A., Hazeldine, D., Kanehisa, M., Kay, L., Lennon, G. G., Lewitter, F. I., Linder, C. R., Luetzenkirchen, A., McCaldon, P., McLeod, M. J., Melone, D. L., Myers, G., Nelson, D., Nial, J.L., Perry, H. M., Rindone, W. P., Sher, L. D., Smith, M. T., Stoesser, G., Swindell, C. D., and Tung, C. S. (1985). "Nucleotide Sequences 1985: A Compilation from the GenBank and EMBL DataLibraries," Vols. 1–4. IRL Press, Oxford, U.K.

Aubert, J. P., Biserte, G., and Loucheux-Lefebvre, M. H. (1976). Carbohydrate-peptide linkage in glycoproteins. *Arch. Biochem. Biophys.* **175**, 410–418.

Auron, P. E., Rindone, W. P., Vary, C. P. H., Celentano, J. J., and Vournakis, J. N. (1982). Computer-aided prediction of RNA secondary structures. *Nucleic Acids Res.* **10**, 403–419.

Avvedimento, V. E., Vogeli, G., Yamada, Y., Maizel, J. V., Pastan, I., and Crombrugghe, B. (1980). Correlation between splicing sites within an intron and their sequence complementarity with U1 RNA. *Cell (Cambridge, Mass.)* **21,** 689–696.

Ayer, D., and Yarus, M. (1986). The context effect does not require a fourth base pair. *Science* **231,** 393–395.

Bach, R., Friedland, P., Brutlag, D. L., and Kedes, L. (1982). MAXAMIZE. A DNA sequencing strategy advisor. *Nucleic Acids Res.* **10,** 295–304.

Bachmair, A., Finley, D., and Varshavsky, A. (1986). In vivo half-life of a protein is a function of its amino-terminal residue. *Science* **234,** 179–186.

Bacon, D. J., and Anderson, W. F. (1986). Multiple sequence alignment. *J. Mol. Biol.* **191,** 153–161.

Bains, W. (1986). MULTAN: A program to align multiple DNA sequences. *Nucleic Acids Res.* **14,** 159–177.

Barlow, D. J., Edwards, M. S., and Thornton, J. M. (1986). Continuous and discontinuous protein antigenic determinants. *Nature (London)* **322,** 747–748.

Bennetzen, J. L., and Hall, B. D. (1982). Codon selection in yeast. *J. Biol. Chem.* **257,** 3026–3031.

Berg, O. G., and von Hippel, P. H. (1987). Selection of DNA binding sites by regulatory proteins: Statistical-mechanical theory and application to operators and promotors. *J. Mol. Biol.* **193,** 723–750.

Berget, S. M. (1984). Are U4 small nuclear ribonucleoproteins involved in polyadenylation? *Nature (London)* **309,** 179–182.

Bernardi, G., and Bernardi, G. (1985). Codon usage and genome composition. *J. Mol. Evol.* **22,** 363–365.

Beyers, T. H., and Waterman, M. S. (1984). Determining all optimal and near-optimal solutions when solving shortest path problems by dynamic programming. *Opt. Res. Q.* **12,** 1381–1384.

Bibb, M. J., Findlay, P. R., and Johnson, M. W. (1984). The relationship between base composition and codon usage in bacterial genes and its use for the simple and reliable identification of protein-coding sequences. *Gene* **30,** 157–166.

Bilofsky, H. S., Burks, C., Fickett, J. W., Goad, W. B., Lewitter, F. I., Rindone, W. P., Swindell, C. D., and Tung, C. S. (1987). Data banks of nucleic acid sequences: GenBank(R). *In* "Introduction to Computing with Protein and Nucleic Acid Sequences" (A. Lesk, ed.). Oxford Univ. Press, London and New York.

Blake, R. D., and Hinds, P. W. (1984). Analysis of codon bias in *E. coli* sequences. *J. Biomol. Struct. Dyn.* **3,** 593–606.

Blundell, T. L., Sibanda, B. L., Sternberg, M. J. E., and Thornton, J. M. (1987). Knowledge-based prediction of protein structures and the design of novel molecules. *Nature (London)* **326,** 347–352.

Borst, P. (1986). How proteins get into microbodies (peroxisomes, glyoxysomes, glycosomes). *Biochim. Biophys. Acta* **866,** 179–203.

Boswell, D. R., and McLachlan, A. D. (1984). Sequence comparison by exponentially-damped alignment. *Nucleic Acids Res.* **12,** 457–464.

Breathnach, R., and Chambon, P. (1981). Organization and expression of eukaryotic split genes coding for proteins. *Annu. Rev. Biochem.* **50,** 349–383.

Breathnach, R., Benoist, C., O'Hare, K., Gannon, F., and Chambon, P. (1978). Ovalbumin gene: Evidence for leader sequence in mRNA and DNA sequences at the exon–intron boundaries. *Proc. Natl. Acad. Sci. U.S.A.* **75,** 4853–4857.

Brendel, V., and Trifonov, E. N. (1984). A computer algorithm for testing potential prokaryotic terminators. *Nucleic Acids Res.* **12,** 4411–4427.

Brendel, V., Hamm, G. H., and Trifonov, E. N. (1986). Terminators of transcription with RNA polymerase from *Escherichia coli:* What they look like and how to find them. *J. Biomol. Struct. Dyn* **3,** 705–723.

Breslauer, K. J., Frank, R., Blöcker, H., and Marky, L. A. (1986). Predicting DNA duplex stability from the base sequence. *Proc. Natl. Acad. Sci. U.S.A.* **83,** 3746–3750.

Briat, J. F., and Chamberlin, M. J. (1984). Identification and characterization of a new transcriptional termination factor from *Escherichia coli. Proc. Natl. Acad. Sci. U.S.A.* **81,** 7373–7377.

Briggs, M. R., Kadonaga, J. T., Bell, S. P., and Tjian, R. (1986). Purification and biochemical characterization of the promoter-specific transcription factor, Sp1. *Science* **234,** 47–52.

Brody, E., and Abelson, J. (1985). The "spliceosome": Yeast pre-messenger RNA associates with a 40S complex in a splicing-dependent reaction. *Science* **228,** 963–967.

Bucher, P., and Trifonov, E. N. (1986). Compilation and analysis of eukaryotic POL II promoter sequences. *Nucl. Acids Res.* **14,** 10009–10026.

Bull, H. B., and Breese, K. (1974). Surface tension of amino acid solutions: A hydrophobicity scale of the amino acid residues. *Arch. Biochem. Biophys.* **161,** 665–670.

Bulmer, M. (1987). Coevolution of codon usage and transfer RNA abundance. *Nature (London)* **325,** 728–730.

Burks, C., Fickett, J. W., Goad, W. B., Kanehisa, M., Lewitter, F. I., Rindone, W. P., Swindell, C. D., Tung, C. S., and Bilofsky, H. S. (1985). The GenBank(R) nucleic acid sequence database. *CABIOS* **1,** 225–233.

Calladine, C. R. (1982). Mechanics of sequence-dependent stacking of bases in B-DNA. *J. Mol. Biol.* **161,** 343–352.

Carter, P. W., Bartkus, J. M., and Calvo, J. M. (1986). Transcription attenuation in *Salmonella typhimurium:* The significance of rare leucine codons in the *leu* leader. *Proc. Natl. Acad. Sci. U.S.A.* **83,** 8127–8131.

Cavener, D. R. (1987). Comparison of the consensus sequence flanking translational start sites in *Drosophila* and vertebrates. *Nucl. Acids Res.* **15,** 1353–1361.

Cellini, A., Parker, R., McMahon, J., Guthrie, C., and Rossi, J. (1986). Activation of a cryptic TACTAAC box in the *Saccharomyces cerevisiae* actin intron. *Mol. Cell. Biol.* **6,** 1571–1578.

Chothia, C. (1974). Hydrophobic bonding and accessible surface area in proteins. *Nature (London)* **248,** 338–339.

Chothia, C. (1976). The nature of the accessible and buried surfaces in proteins. *J. Mol. Biol.* **105,** 1–14.

Chothia, C. (1984). Principles that determine the structure of proteins. *Annu. Rev. Biochem.* **53,** 537–572.

Chou, P. Y., and Fasman, G. D. (1974a). Conformational parameters for amino acids in helical, β-sheet, and random coil regions calculated from proteins. *Biochemistry* **13,** 211–222.

Chou, P. Y., and Fasman, G. D. (1974b). Prediction of protein conformation. *Biochemistry* **13,** 222–245.

Chou, P. Y., and Fasman, G. D. (1978a). Empirical predictions of protein conformations. *Annu. Rev. Biochem.* **47,** 251–276.

Chou, P. Y., and Fasman, G. D. (1978b). Prediction of the secondary structure of proteins from their amino acid sequence. *Adv. Enzymol. Relat. Areas Mol. Biol.* **47,** 45–148.

Ciliberto, G., Raugel, G., Costanzo, F., Dente, L., and Cortese, R. (1983). Common and interchangeable elements in promoters of genes transcribed by RNA polymerase III. *Cell (Cambridge, Mass.)* **32,** 725–733.

Claverie, J. M., and Sauvaget, I. (1985). A new protein sequence data bank. *Nature (London)* **318,** 19.

Cohen, F. E., Abarbanel, R. M., Kuntz, I. D., and Fletterick, R. J. (1983). Secondary structure assignment for α/β proteins by a combinatorial approach. *Biochemistry* **22,** 4894–4904.

Cohen, F. E., Abarbanel, R. M., Kuntz, I. D., and Fletterick, R. J. (1986). Turn prediction in proteins using a pattern-matching approach. *Biochemistry* **25,** 266–275.

Cowing, D. W., Bardwell, J. C. A., Craig, E. A., Woolford, C., Hendrix, R. W., and Gross, C. A. (1985). Consensus sequence for *Escherichia coli* heat shock gene promoters. *Proc. Natl. Acad. Sci. U.S.A.* **82,** 2679–2683.

Craik, C. S., Sprang, S., Fletterick, R., and Rutter, W. J. (1982). Intron–exon splice junctions map at protein surfaces. *Nature (London)* **299,** 180–182.

Creighton, T. E. (1983). "Proteins: Structures and Molecular Properties." Freeman, New York.

Davey, J., Dimmock, N. J., and Colman, A. (1985). Identification of the sequence responsible for the nuclear accumulation of the influenza virus nucleoprotein in *Xenopus* oocytes. *Cell (Cambridge, Mass.)* **40,** 667–675.

Davison, D. (1985). Sequence similarity ('homology') searching for molecular biologists. *Bull. Math. Biol.* **47,** 437–474.

Dayhoff, M. O., ed. (1972). "Atlas of Protein Sequence and Structure," Vol. 5. Natl. Biomed. Res. Found., Washington, D.C.

Dean, G. E., MacNab, R. M., Stafer, J., Matsumura, P., and Burks, C. (1984). Gene sequence and predicted amino acid sequence of the motA protein, a membrane-associated protein required for flagellar rotation in *Escherichia coli. J. Bacteriol.* **159,** 991–999.

de Crombrugghe, B., Busby, S., and Buc, H. (1984). Cyclic AMP receptor protein: Role in transcription activation. *Science* **224,** 831–838.

Deisenhofer, J., Epp, O., Miki, K., Huber, R., and Michel, H. (1985). Structure of the protein subunits in the photosynthetic reaction centre of *Rhodopseudomonas viridis* at 3Å resolution. *Nature (London)* **318,** 618–624.

Deuschle, U., Kammerer, W., Gentz, R., and Bujard, H. (1986). Promoters of *Escherichia coli:* A hierarchy of in vivo strength indicates alternate structures. *EMBO J.* **5,** 2987–2994.

De Wachter, R. (1979). Do eukaryotic mRNA 5′ noncoding sequences base-pair with the 18S ribosomal RNA 3′ terminus? *Nucleic Acids Res.* **7,** 2045–2054.

Dickerson, R. E. (1983). Base sequence and helical structure variation in B and A DNA. *J. Mol. Biol.* **166,** 419–441.

Domdey, H., Apostol, B., Lin, R. J., Newman, A., Brody, E., and Abelson, J. (1984). Lariat structures are in vivo intermediates in yeast pre-mRNA splicing. *Cell (Cambridge, Mass.)* **39,** 611–621.

Douglas, M. G., McCammon, M. T., and Vassarotti, A. (1986). Targeting proteins into mitochondria. *Microbiol. Rev.* **50,** 166–178.

Douthart, R. J., Thomas, J. J., Rosier, S. D., Schmaltz, J. E., and West, J. W. (1986). Cloning simulation in the CAGE(tm) environment. *Nucleic Acids Res.* **14,** 285–297.

Drew, H. R., Wing, R. N., Takano, T., Broka, C., Tanaka, S., Itakura, I., and Dickerson, R. E. (1981). Structure of a B-DNA dodecamer: Conformation and dynamics. *Proc. Natl. Acad. Sci. U.S.A.* **78,** 2179–2183.

Duggleby, R. G., Kinns, H., and Rood, J. I. (1981). A computer program for determining the size of DNA restriction fragments. *Anal. Biochem.* **110,** 49–55.

Dumas, J. P., and Ninio, J. (1982). Efficient algorithm for folding and comparing nucleic acid sequences. *Nucleic Acids Res.* **10,** 197–206.

Durand, R., and Bregegere, F. (1984). An efficient program to construct restriction maps from experimental data with realistic error levels. *Nucleic Acids Res.* **12,** 703–716.

Dynan, W. S., and Tjian, R. (1985). Control of eukaryotic messenger RNA synthesis by sequence-specific DNA-binding proteins. *Nature (London)* **316,** 774–778.

Eisenberg, D. (1984). Three-dimensional structure of membrane and surface proteins. *Annu. Rev. Biochem.* **53,** 595–623.

Eisenberg, D., and McLachlan, A.D. (1986). Solvation energy in protein folding and binding. *Nature (London)* **319,** 199–203.

Eisenberg, D., Schwarz, E., Komaromy, M., and Wall, R. (1984a). Analysis of membrane and surface protein sequences with the hydrophobic moment plot. *J. Mol. Biol.* **179,** 125–142.

Eisenberg, D., Weiss, R. M., and Terwilliger, T. C. (1984b). The hydrophobic moment detects periodicity in protein hydrophobicity. *Proc. Natl. Acad. Sci. U.S.A.* **81,** 140–144.

Engelman, D. M., and Steitz, T. A. (1981). The spontaneous insertion of proteins into and across membranes: The helical hairpin hypothesis. *Cell (Cambridge, Mass.)* **23,** 411–422.

Engelman, D. M., Steitz, T. A., and Goldman, A. (1986). Identifying nonpolar transbilayer helices in amino acid sequences of membrane proteins. *Annu. Rev. Biophys. Biophys. Chem.* **15,** 321–353.

Eperon, L. P., Estibeiro, J. P., and Eperon, I. C. (1986). The role of nucleotide sequences in splice site selection in eukaryotic pre-messenger RNA. *Nature (London)* **324,** 280–282.

Fanning, D. W., Smith, J. A., and Rose, G. D. (1986). Molecular cartography of globular proteins with application to antigenic sites. *Biopolymers* **25,** 863–883.

Farnham, P. J., and Platt, T. (1980). A model for transcription termination suggested by studies on the *trp* attenuator in vitro using base analogs. *Cell (Cambridge, Mass.)* **20,** 739–748.

Fauchère, J. L., and Pliska, V. (1983). Hydrophobicity parameters π of amino acid side chains from the partitioning of *N*-acetyl-amino-acid amides. *Eur. J. Med. Chem. —Chim. Ther.* **18,** 369–375.

Feng, D. F., Johnson, M. S., and Doolittle, R. F. (1985). Aligning amino acid sequences: Comparison of commonly used methods. *J. Mol. Evol.* **21,** 112–125.

Fickett, J. W. (1982). Recognition of protein coding regions in DNA sequences. *Nucleic Acids Res.* **10,** 5303–5318.

Finer-Moore, J., and Stroud, R. M. (1984). Amphipathic analysis and possible formation of the ion channel in an acetylcoline receptor. *Proc. Natl. Acad. Sci. U.S.A.* **81,** 155–159.

Fisher, R. F., Das, A., Kolter, R., Winkler, M. E., and Yanofsky, C. (1985). Analysis of the requirements for transcription pausing in the tryptophan operon. *J. Mol. Biol.* **182,** 397–409.

Fitch, W. M. (1974). The large extent of putative secondary nucleic acid structure in random nucleotide sequences or amino acid derived messenger-RNA. *J. Mol. Evol.* **3,** 279–291.

Fitch, W. M. (1983). Calculating the expected frequencies of potential secondary structure in nucleic acids as a function of stem length, loop size, base composition and nearest neighbor frequencies. *Nucleic Acids Res.* **11,** 4655–4663.

Fitch, W. M., and Smith, T. F. (1983). Optimal sequence alignments. *Proc. Natl. Acad. Sci. U.S.A.* **80,** 1382–1386.

Fitzgerald, M., and Shenk, T. (1981). The sequence 5′-AAUAAA-3′ forms part of the recognition site for polyadenylation of late SV40 mRNAs. *Cell (Cambridge, Mass.)* **24,** 251–260.

Fixman, M., and Freire, J. J. (1977). Theory of DNA melting curves. *Biopolymers* **16,** 2693–2704.

Flinta, C., von Heijne, G., and Johansson, J. (1983). Helical sidedness and the distribution of polar residues in trans-membrane helices. *J. Mol. Biol.* **168,** 193–196.

Flinta, C., Persson, B., Jörnvall, H., and von Heijne, G. (1986). Sequence determinants of N-terminal protein processing. *Eur. J. Biochem.* **154,** 193–196.

Foley, B. T., Nelson, D., Smith, M. T., and Burks, C. (1986). Cross-section of the GenBank database. *Trends Genetics* (September), 233–238.

Freier, S. M., Kierzek, R., Jaeger, J. A., Sugimoto, N., Caruthers, M. H., Neilson, T., and Turner, D. H. (1986). Improved free-energy parameters for predictions of RNA duplex stability. *Proc. Natl. Acad. Sci. U.S.A.* **83,** 9373–9377.

Frömmelt, C. (1984). The apolar surface area of amino acids and its empirical correlation with hydrophobic free energy. *J. Theor. Biol.* **111,** 247–260.

Furdon, P. J., and Kole, R. (1986). Inhibition of splicing but not cleavage at the 5′ splice site by truncating human β-globin pre-mRNA. *Proc. Natl. Acad. Sci. U.S.A.* **83,** 927–931.

Galas, D. J., Eggert, M., and Waterman, M. S. (1985). Rigorous pattern-recognition methods for DNA sequences. Analysis of promoter sequences from *Escherichia coli. J. Mol. Biol.* **186,** 117–128.

Ganoza, M. C., Kofoid, E. C., Marlière, P., and Louis, B. G. (1987). Potential secondary structure at translation-initiation sites. *Nucl. Acids Res.* **15,** 345–360.

Garnier, J., Osguthorpe, D. J., and Robson, B. (1978). Analysis of the accuracy and implications of simple methods for predicting the secondary structure of globular proteins. *J. Mol. Biol.* **120,** 97–120.

Garratt, R. C., Taylor, W. R., and Thornton, J. M. (1985). The influence of tertiary structure on secondary structure prediction. *FEBS Lett.* **188,** 59–62.

George, D. G., Barker, W. C., and Hunt, L. T. (1986). The protein identification resource (PIR). *Nucleic Acids Res.* **14,** 11–15.

Gibbs, A. J., and McIntyre, G. A. (1970). The diagram, a method for comparing sequences. *Eur. J. Biochem.* **16,** 1–11.

Gilbert, W. (1985). Genes-in-pieces revisited. *Science* **228,** 823–824.

Gilbert, W., Marchionni, M., and McKnight, G. (1986). On the antiquity of introns. *Cell (Cambridge, Mass.)* **46,** 151–154.

Gō, M. (1981). Correlation of DNA exonic regions with protein structural units in haemoglobin. *Nature (London)* **291,** 90–92.

Gō, M. (1983a). Modular structural units, exons, and function in chicken lysozyme. *Proc. Natl. Acad. Sci. U.S.A.* **80,** 1964–1968.

Gō, M. (1983b). Theoretical studies of protein folding. *Annu. Rev. Biophys. Bioeng.* **12,** 183–210.

Goad, W. B. (1986). Computational analysis of genetic sequences. *Annu. Rev. Biophys. Biophys. Chem.* **15,** 79–95.

Goad, W. B., and Kanehisa, M. I. (1982). Pattern recognition in nucleic acid sequences. I. A. general method for finding local homologies and symmetries. *Nucleic Acids Res.* **10,** 247–263.

Gotoh, O. (1986). Alignment of three biological sequences with an efficient traceback procedure. *J. Theor. Biol.* **121,** 327–337.

Grantham, R. (1978). Viral, prokaryote and eukaryote genes contrasted by mRNA sequence indexes. *FEBS Lett.* **95,** 1–11.

Grantham, R., Gautier, C., Gouy, M., Mercier, R., and Pavé, A. (1980). Codon catalog usage and the genome hypothesis. *Nucleic Acids Res.* **8,** r49–r60.

Gribskov, M., Devereux, J., and Burgess, R. R. (1984). The codon preference plot: Graphic analysis of protein coding sequences and prediction of gene expression. *Nucleic Acids Res.* **12,** 539–549.

Gribskov, M., McLachlan, A. D., and Eisenberg, D. (1987). Profile analysis: Detection of distantly related proteins. *Proc. Natl. Acad. Sci. U.S.A.* (in press).

Grosjean, H., Sankoff, D., Min Jou, W., Fiers, W., and Cedergren, R. J. (1978). Bacteriophage MS2 RNA: A correlation between the stability of the codon-anticodon interaction and the choice of code words. *J. Mol. Evol.* **12,** 113–119.

Grymes, R. A., Travers, P., and Engelberg, A. (1986). GEL—A computer tool for DNA sequencing projects. *Nucleic Acids Res.* **14,** 87–99.

Guy, H. R. (1984). A structural model of the acetylcholine receptor channel based on partition energy and helix packing calculations. *Biophys. J.* **45,** 249–261.

Guy, H. R. (1985). Amino acid side-chain partition energies and distribution of residues in soluble proteins. *Biophys. J.* **47,** 61–70.

Guy, H. R., and Seetharamulu, P. (1986). Molecular model of the action potential sodium channel. *Proc. Natl. Acad. Sci. U.S.A.* **83,** 508–512.

Hagerman, P. J. (1986). Sequence-directed curvature of DNA. *Nature (London)* **321,** 449–450.

Hamed, M. M., Robinson, R. M., and Mattice, W. L. (1983). Behavior of amphipathic helices on analysis via matrix methods, with application to glucagon, secretin, and vasoactive intestinal peptide. *Biopolymers* **22,** 1003–1021.

Harrison, S. C. (1986). Fingers and DNA half-turns. *Nature (London)* **322,** 597–598.

Hawley, D. K., and McClure, W. R. (1983). Compilation and analysis of *Escherichia coli* promoter DNA sequences. *Nucleic Acids Res.* **11,** 2237–2255.

Henikoff, S., and Cohen, E. H. (1984). Sequences responsible for transcription termination on a gene segment in *Saccharomyces cerevisiae. Mol. Cell. Biol.* **4,** 1515–1520.

Hersko, A., Heller, H., Eytan, E., Kaklij, G., and Rose, I. A. (1984). Role of the α-amino group of protein in ubiquitin-mediated protein breakdown. *Proc. Natl. Acad. Sci U.S.A.* **81,** 7021–7025.

Ho, P. S., Ellison, M. J., Quigley, G. J., and Rich, A. (1986). A computer aided thermodynamic approach for predicting the formation of Z-DNA in naturally occurring sequences. *EMBO J.* **5,** 2737–2744.

Holley, R. W., Apgar, J., Everett, G. A., Madison, J. T., Marquisee, M., Merrill, S. H., Penswick, J. R., and Zamir, A. (1965). Structure of a ribonucleic acid. *Science* **147,** 1462–1465.

Holm, L. (1986). Codon usage and gene expression. *Nucleic Acids Res.* **14,** 3075–3087.

Hopp, T. P., and Woods, K. R. (1981). Prediction of protein antigenic determinants from amino acid sequences. *Proc. Natl. Acad. Sci. U.S.A.* **78,** 3824–3828.

Houghton, M., Eaton, M. A. W., Stewart, A. G., Smith, J. C., Doel, S. M., Catlin, G. H., Lewis, H. M., Patel, T. P., Emtage, J. S., Carey, N. H., and Porter, A. G. (1980). The complete amino acid sequence of human fibroblast interferon as deduced using synthetic oligodeoxyribonucleotide primers of reverse transcriptase. *Nucleic Acids Res.* **8,** 2885–2893.

Hubbard, S. C., and Ivatt, R. J. (1981). Synthesis and processing of asparagine-linked oligosaccharides. *Annu. Rev. Biochem.* **50,** 555–583.

Hunt, L. T., and Dayhoff, M. O. (1970). The occurrence in proteins of the tripeptides Asn-X-Ser and Asn-X-Thr and of bound carbohydrate. *Biochem. Biophys. Res. Commun.* **39,** 757–765.

Ikemura, T. (1981a). Correlation between the abundance of *Escherichia coli* transfer RNAs and the occurrence of the respective codons in its protein genes. *J. Mol. Biol.* **146,** 1–21.

Ikemura, T. (1981b). Correlation between the abundance of *Escherichia coli* transfer RNAs and the occurrence of the respective codons in its protein genes. *J. Mol. Biol.* **151,** 389–409.

Ikemura, T. (1985). Codon usage and tRNA content in unicellular and multicellular organisms. *Mol. Biol. Evol.* **2,** 13–34.

Itakura, K., Rossi, J. J., and Wallace, R. B. (1984). Synthesis and use of synthetic oligonucleotides. *Annu. Rev. Biochem.* **53,** 323–356.

Janin, J. (1979). Surface and inside volumes in globular proteins. *Nature (London)* **277,** 491–492.

Jungck, J. R., and Friedman, R. M. (1984). Mathematical tools for molecular genetics data: An annotated bibliography. *Bull. Math. Biol.* **46,** 699–744.

Jurka, J., and Savageau, M. A. (1985). Gene density over the chromosome of *Escherichia coli:* Frequency distribution, spatial clustering, and symmetry. *J. Bacteriol.* **163,** 806–811.

Kabsch, W., and Sander, C. (1983). How good are predictions of protein secondary structure? *FEBS Lett.* **155,** 179–182.

Kaiser, E. T., and Kezdy, F. J. (1984). Amphiphilic secondary structure: Design of peptide hormones. *Science* **223,** 249–255.

Kammerer, W., Deuschle, U., Gentz, R., and Bujard, H. (1986). Functional dissection of *Escherichia coli* promoters: Information in the transcribed region is involved in late steps of the overall process. *EMBO J.* **5,** 2995–3000.

Kanehisa, M. I. (1984). Use of a statistical criteria for screening potential homologies in nucleic acid sequences. *Nucleic Acids Res.* **12,** 203–213.

Kanehisa, M. I., and Goad, W. B. (1982). Pattern recognition in nucleic acid sequences. II. An efficient method for finding locally stable secondary structures. *Nucleic Acids Res.* **10,** 265–278.

Karlin, S., and Ghandour, G. (1985). Multiple-alphabet amino acid sequence comparison of the immunoglobulin κ-chain constant domain. *Proc. Natl. Acad. Sci. U.S.A.* **82,** 8597–8601.

Karlin, S., Ghandour, G., Ost, F., Tavare, S., and Korn, L. J. (1983). New approaches for computer analysis of nucleic acid sequences. *Proc. Natl. Acad. Sci. U.S.A.* **80,** 5660–5664.

Karlin-Neumann, G. A., and Tobin, E. M. (1986). Transit peptides of nuclear-encoded chloroplast proteins share a common amino acid framework. *EMBO J.* **5,** 9–13.

Karplus, P. A., and Schultz, G. E. (1985). Prediction of chain flexibility in proteins. *Naturwissenschaften* **72,** 212–213.

Keller, E. B., and Noon, W. A. (1984). Intron splicing: A conserved internal signal in introns of animal pre-mRNAs. *Proc. Natl. Acad. Sci. U.S.A.* **81,** 7417–7420.

Keller, W. (1984). The RNA lariat: A new ring to the splicing of mRNA precursors. *Cell (Cambridge, Mass.)* **39,** 423–425.

Klein, P., and DeLisi, C. (1986). Prediction of protein structural class from the amino acid sequence. *Biopolymers* **25,** 1659–1672.

Klein, P., Kanehisa, M., and DeLisi, C. (1984). Prediction of protein function from sequence properties. *Biochim. Biophys. Acta* **787,** 221–226.

Klein, P., Kanehisa, M., and DeLisi, C. (1985). The detection and classification of membrane-spanning proteins. *Biochim. Biophys. Acta* **815,** 468–476.

Klein, P., Jacquez, J. A., and DeLisi, C. (1986). Prediction of protein function by discriminant analysis. *Math. Biosci.* **81,** 177–190.

Konopka, A. K., Reiter, J., Jung, M., Zarling, D. A., and Jovin, T. M. (1985). Concordance of experimentally mapped or predicted Z-DNA sites with positions of selected alternating purine–pyrimidine tracts. *Nucleic Acids Res.* **13,** 1683–1701.

Koo, H. S., Wu, H. M., and Crothers, D. M. (1986). DNA bending at adenine–thymine tracts. *Nature (London)* **320,** 501–506.

Korn, L. J., and Queen, C. (1984). Analysis of biological sequences on small computers. *DNA* **3,** 421–436.

Kovacs, B. J., and Butterworth, P. H. W. (1986). The effect of changing the distance between the TATA-box and cap site by up to three base pairs on the selection of the transcriptional start site of a cloned eukaryotic gene *in vitro* and *in vivo. Nucleic Acids Res.* **14,** 2429–2442.

Kozak, M. (1981). Possible role of flanking nucleotides in recognition of the AUG initiator codon by eukaryotic ribosomes. *Nucleic Acids Res.* **9,** 5233–5252.

Kozak, M. (1983). Comparison of initiation of protein synthesis in prokaryotes, eukaryotes, and organelles. *Microbiol. Rev.* **47,** 1–45.

Kozak, M. (1984a). Point mutations close to the AUG initiator codon affect the efficiency of translation of rat preproinsulin in vitro. *Nature (London)* **308,** 241–246.

Kozak, M. (1984b). Compilation and analysis of sequences upstream from the translational start site in eukaryotic mRNAs. *Nucleic Acids Res.* **12,** 857–872.

Kozak, M. (1986a). Point mutations define a sequence flanking the AUG initiator codon that modulates translation by eukaryotic ribosomes. *Cell (Cambridge, Mass.)* **44,** 283–292.

Kozak, M. (1986b). Influences of mRNA secondary structure on initiation by eukaryotic ribosomes. *Proc. Natl. Acad. Sci. U.S.A.* **83,** 2850–2854.

Kuhn, L. A., and Leigh, J. S. (1985). A statistical technique for predicting membrane protein structure. *Biochim. Biophys. Acta* **828,** 351–361.

Kyte, J., and Doolittle, R. F. (1982). A simple method for displaying the hydropathic character of a protein. *J. Mol. Biol.* **157,** 105–132.

Langford, C. J., and Gallwitz, D. (1983). Evidence for an intron-contained sequence required for the splicing of yeast RNA polymerase II transcripts. *Cell (Cambridge, Mass.)* **33,** 519–527.

Lathe, R. (1985). Synthetic oligonucleotide probes deduced from amino acid sequence data. Theoretical and practical considerations. *J. Mol. Biol.* **183,** 1–12.

Lennon, G. G., and Nussinov, R. (1985). Eukaryotic oligomer frequencies are correlated with certain DNA helical parameters. *J. Theor. Biol.* **22,** 427–433.

Lenstra, J. A. (1977). Evaluation of secondary structure predictions in proteins. *Biochim. Biophys. Acta* **491,** 333–338.

Lerner, M. R., Boyle, J. A., Mount, S. M., Wolin, S. L., and Steitz, J. A. (1980). Are snRNPs involved in splicing? *Nature (London)* **283,** 220–224.

Lesk, A. M. (1985). Coordination of sequence data. *Nature (London)* **314,** 318–319.

Levin, J. M., Robson, B., and Garnier, J. (1986). An algorithm for secondary structure determination in proteins based on sequence similarity. *FEBS Lett.* **205,** 303–308.

Levitt, M. (1978). Conformational preferences of amino acids in globular proteins. *Biochemistry* **17,** 4277–4285.

Lewin, R. (1986). DNA databases are swamped. *Science* **232,** 1599.

Lewis, R. M. (1986). PROBFIND: A computer program for selecting oligonucleotide probes from peptide sequences. *Nucleic Acids Res.* **14,** 567–570.

Liljenström, H., von Heijne, G., Blomberg, C., and Johansson, J. (1985). The tRNA cycle and its relation to the rate of protein synthesis. *Eur. Biophys. J.* **12,** 115–119.

Lim, V. I. (1974a). Structural principles of the globular organization of protein chains. A stereochemical theory of globular protein secondary structure. *J. Mol. Biol.* **88,** 857–872.

Lim, V. I. (1974b). Algorithms for prediction of α-helical and β-structural regions in globular proteins. *J. Mol. Biol.* **88,** 873–894.

Lipman, D. J., and Pearson, W. R. (1985). Rapid and sensitive protein similarity searches. *Science* **227,** 1435–1441.

Lipman, D. J., and Wilbur, W. J. (1983). Contextual constraints on synonymous codon choice. *J. Mol. Biol.* **163,** 363–376.

Lipman, D. J., Wilbur, W. J., Smith, T. F., and Waterman, M. S. (1984). On the statistical significance of nucleic acid similarities. *Nucleic Acids Res.* **12,** 215–226.

Lockard, R. E., Currey, K., Browner, M., Lawrence, C., and Maizel, J. (1986). Secondary structure model for mouse β major globin mRNA derived from enzymatic digestion data, comparative sequence and computer analysis. *Nucl. Acids Res.* **14,** 5827–5841.

Loomis, W. F., and Gilpin, M. E. (1986). Multigene families and vestigial sequences. *Proc. Natl. Acad. Sci. U.S.A.* **83,** 2143–2147.

Lütcke, H. A., Chow, K. C., Mickel, F. S., Moss, K. A., Kern, H. F., and Scheele, G. A. (1987). Selection of AUG initiation codons differs in plants and animals. *EMBO J.* **6,** 43–48.

McClure, W. R., Hawley, D. K., Youderian, P., and Susskind, M. M. (1983). DNA determinants of promoter selectivity in *Escherichia coli*. *Cold Spring Harbor Symp. Quant. Biol.* **47,** 477–481.

McDevitt, M. A., Hart, R. P., Wong, W. W., and Nevins, J. R. (1986). Sequences capable of restoring poly(A) site function define two distinct downstream elements. *EMBO J.* **5,** 2907–2913.

McGeoch, D. J. (1985). On the predictive recognition of signal peptide sequences. *Virus Res.* **3,** 271–286.

McLachlan, A. D. (1971). Test for comparing related amino-acid sequences. Cytochrome *c* and cytochrome *c*551. *J. Mol. Biol.* **61,** 409–424.

McLachlan, A. D., Staden, R., and Boswell, D. R. (1984). A method for measuring the non-random bias of codon usage. *Nucleic Acids Res.* **12,** 9567–9575.

McLauchlan, J., Gaffney, D., Whitton, J. L., and Clements, J. B. (1985). The consensus sequence YGTGTTYY located downstream from the AATAAA signal is required for efficient formation of mRNA 3′ termini. *Nucleic Acids Res.* **13,** 1347–1368.

McLean, M. J., Blaho, J. A., Kilpatrick, M. W., and Wells, R. D. (1986). Consecutive AT pairs can adopt a left-handed DNA structure. *Proc. Natl. Acad. Sci. U.S.A.* **83,** 5884–5888.

Maizel, J. V., and Lenk, R. P. (1981). Enhanced graphic matrix analysis of nucleic acid and protein sequences. *Proc. Natl. Acad. Sci. U.S.A.* **78,** 7665–7669.

Manabe, T. (1981). Theory of regulation by the attenuation mechanism: Stochastic model for the attenuation of the *Escherichia coli* tryptophan operon. *J. Theor. Biol.* **91,** 527–544.

Manavalan, P., and Ponnuswamy, P. K. (1978). Hydrophobic character of amino acid residues in globular proteins. *Nature (London)* **275,** 673–674.

Marck, C. (1986). Fast analysis of DNA and protein sequence on Apple IIe: Restriction sites search, alignment of short sequences and dot matrix analysis. *Nucleic Acids Res.* **14,** 583–590.

Maroun, L. E., Degner, M., Precup, J. W., and Franciskovich, P. P. (1986). Eukaryotic mRNA 5′-leader sequences have dual regions of complementarity to the 3′-terminus of 18S rRNA. *J. Theor. Biol.* **119,** 85–98.

Martin, F. H., and Castro, M. M. (1985). Base pairing involving deoxyinosine: Implications for probe design. *Nucleic Acids Res.* **13,** 8927–8938.

Martinez, H. M. (1983). An efficient method for finding repeats in molecular sequences. *Nucleic Acids Res.* **11,** 4629–4634.

Martinez, H. M. (1984). An RNA folding rule. *Nucleic Acids Res.* **12,** 323–334.

Maruyama, T., Gojobori, T., Aota, S., and Ikemura, T. (1986). Codon usage tabulated from the GenBank genetic sequence data. *Nucleic Acids Res.* **14,** r151–r189.

Mason, P. J., Jones, M. B., Elkington, J. A., and Williams, J. G. (1985). Polyadenylation of the *Xenopus* β1 globin mRNA at a downstream minor site in the absence of the major site and utilization of an AAUACA polyadenylation signal. *EMBO J.* **4,** 205–211.

Maxam, A. M., and Gilbert, W. (1977). A new method for sequencing DNA. *Proc. Natl. Acad. Sci. U.S.A.* **74,** 560–564.

Mengeritsky, G., and Trifonov, E. N. (1983). Nucleotide sequence-directed mapping of the nucleosome. *Nucleic Acids Res.* **11,** 3833–3851.

Michel, C. J. (1986). New statistical approach to discriminate between protein coding and non-coding regions in DNA sequences and its evaluation. *J. Theor. Biol.* **120,** 223–236.

Miller, J., McLachlan, A., and Klug, A. (1985). Repetitive zinc-binding domains in the protein transcription factor IIIA from *Xenopus* oocytes. *EMBO J.* **4,** 1609–1614.

Mironov, A. A., Dyakonova, L. P., and Kister, A. E. (1985). A kinetic approach to the prediction of RNA secondary structures. *J. Biomol. Struct. Dyn.* **2,** 953–962.

Miyazawa, S., and Jernigan, R. L. (1985). Estimation of effective interresidue contact energies from protein crystal structures: Quasi-chemical approximation. *Macromolecules* **18,** 534–552.

Moreland, R. B., Nam, H. G., Hereford, L. M., and Fried, H. M. (1985). Identification of a nuclear localization signal of a yeast ribosomal protein. *Proc. Natl. Acad. Sci. U.S.A.* **82,** 6561–6565.

Mount, D. W. (1984). Modeling RNA structure. *Bio/Technology,* September, pp. 791–795.

Mount, D. W. (1985). Computer analysis of sequence, structure and function of biological macromolecules. *BioTechniques,* March/April, pp. 102–112.

Mount, S. M. (1982). A catalogue of splice junction sequences. *Nucleic Acids Res.* **10,** 459–472.

Mulligan, M. E., and McClure, W. R. (1986). Analysis of the occurrence of promoter-sites in DNA. *Nucleic Acids Res.* **14,** 109–126.

Mulligan, M. E., Hawley, D. K., Entriken, R., and McClure, W. R. (1984). *Escherichia coli* promoter sequences predict in vitro RNA polymerase selectivity. *Nucleic Acids Res.* **12,** 789–800.

Murata, M., Richardson, J. S., and Sussman, J. L. (1985). Simultaneous comparison of three protein sequences. *Proc. Natl. Acad. Sci. U.S.A.* **82,** 3073–3077.

Myers, R. M., Tilly, K., and Maniatis, T. (1986). Fine structure genetic analysis of a β-globin promoter. *Science* **232,** 613–618.

Nakashima, H., Nishikawa, K., and Ooi, T. (1986). The folding type of a protein is relevant to the amino acid composition. *J. Biochem. (Tokyo)* **99,** 153–162.

Nakata, K., Kanehisa, M., and DeLisi, C. (1985). Prediction of splice junctions in mRNA sequences. *Nucleic Acids Res.* **13,** 5327–5340.

Needleman, S. B., and Wunsch, C. D. (1970). A general method applicable to the search for similarities in the amino acid sequence of two proteins. *J. Mol. Biol.* **48,** 443–453.

Nishikawa, K. (1983). Assessment of secondary-structure prediction of proteins. *Biochim. Biophys. Acta* **748,** 285–299.

Nishikawa, K., and Ooi, T. (1986). Amino acid sequence homology applied to the prediction of protein secondary structures, and joint prediction with existing methods. *Biochim. Biophys. Acta* **871,** 45–54.

Nishikawa, K., Kubota, Y., and Ooi, T. (1983a). Classification of proteins into groups based on amino acid composition and other characters. I. *J. Biochem. (Tokyo)* **94,** 981–995.

Nishikawa, K., Kubota, Y., and Ooi, T. (1983b). Classification of proteins into groups based on amino acid composition and other characters. II. *J. Biochem. (Tokyo)* **94,** 997–1007.

Nolan, G. P., Maina, C. V., and Szalay, A. A. (1984). Plasmid mapping computer program. *Nucleic Acids Res.* **12,** 717–729.

Novotny, J., Handschumacher, M., Haber, E., Bruccoleri, R. E., Carlson, W. B., Fanning, D. W., Smith, J. A., and Rose, G. D. (1986). Antigenic determinants in proteins coincide with surface regions accessible to large probes (antibody domains). *Proc. Natl. Acad. Sci. U.S.A.* **83,** 226–230.

Nozaki, Y., and Tanford, C. (1971). The solubility of amino acids and two glycine peptides in aqueous ethanol and dioxane solutions. *J. Biol. Chem.* **246,** 2211–2217.

Nussinov, R. (1982). RNA folding is unaffected by the nonrandom degenerate codon choice. *Biochim. Biophys. Acta* **698,** 111–115.

Nussinov, R. (1986). Sequence signals which may be required for efficient formation of mRNA 3′ termini. *Nucleic Acids Res.* **14,** 3557–3571.

Nussinov, R., and Jacobson, A. B. (1980). Fast algorithm for predicting the secondary structure of single-stranded DNA. *Proc. Natl. Acad. Sci. U.S.A.* **77,** 6309–6313.

Nussinov, R., and Lennon, G. G. (1984). Structural features are as important as sequence homologies in drosophilia heat shock gene upstream regions. *J. Mol. Evol.* **20,** 106–110.

Nussinov, R., Shapiro, B., Lipkin, L. E., and Maizel, J. V. (1984a). Enhancer elements share local homologous twist-angle variations with a helical periodicity. *Biochim. Biophys. Acta* **783,** 246–257.

Nussinov, R., Shapiro, B., Lipkin, L. E., and Maizel, J. V. (1984b). DNAase I hypersensitive sites may be correlated with genomic regions of large structural variation. *J. Mol. Biol.* **177,** 591–607.

Nussinov, R., Owens, J., and Maizel, J. V. (1986). Sequence signals in eukaryotic upstream regions. *Biochim. Biophys. Acta* **866,** 109–119.

Ohlendorf, D. H., and Matthews, B. W. (1983). Structural studies of protein–nucleic acid interactions. *Annu. Rev. Biophys. Bioeng.* **12,** 259–284.

Ohshima, Y., Itoh, M., Okada, N., and Miyata, T. (1981). Novel models for RNA splicing that involve a small nuclear RNA. *Proc. Natl. Acad. Sci. U.S.A.* **78,** 4471–4474.

Orcutt, B. C., George, D. G., and Dayhoff, M. O. (1983). Protein and nucleic acid sequence database systems. *Annu. Rev. Biophys. Bioeng.* **12,** 419–441.

Pabo, C. O., and Sauer, R. T. (1984). Protein–DNA recognition. *Annu. Rev. Biochem.* **53,** 293–321.

Paolella, G., and Russo, T. (1985). A microcomputer program for the identification of tRNA genes. *CABIOS* **1,** 149–151.

Papanicolaou, C., Gouy, M., and Ninio, J. (1984). An energy model that predicts the correct folding of both the tRNA and the 5S RNA molecules. *Nucleic Acids Res.* **12,** 31–44.

Parker, J., Johnston, T. C., Boria, P. T., Holtz, G., Remaut, E., and Fiers, W. (1983). Codon usage and mistranslation. *J. Biol. Chem.* **258,** 10007–10012.

Parker, J. M. R., Guo, D., and Hodges, R. S. (1986). New hydrophilicity scale derived from high-performance liquid chromatography peptide retention data: Correlation of predicted surface residues with antigenicity and X-ray-derived accessible sites. *Biochemistry* **25,** 5425–5432.

Pattus, F., Heitz, F., Martinez, C., Provencher, S. W., and Lazdunski, C. (1985). Secondary structure of the pore-forming colicin A and its C-terminal fragment. *Eur. J. Biochem.* **152,** 681–689.

Paul, C., and Rosenbusch, J. P. (1985). Folding patterns of porin and bacteriorhodopsin. *EMBO J.* **4,** 1593–1597.

Pearson, W. R. (1982). Automatic construction of restriction site maps. *Nucleic Acids Res.* **10,** 217–227.

Pelham, H. (1985). Activation of heat-shock genes in eukaryotes. *Trends Genet.* January, pp. 31–35.

Persson, B., Flinta, C., von Heijne, G., and Jörnvall, H. (1985). Structures of N-terminally acetylated proteins. *Eur. J. Biochem.* **152,** 523–527.

Pipas, J. M., and McMahon, J. E. (1975). Method for predicting RNA secondary structure. *Proc. Natl. Acad. Sci. U.S.A.* **72,** 2017–2021.

Platt, T. (1986). Transcription termination and the regulation of gene expression. *Annu. Rev. Biochem.* **55,** 339–372.

Pollack, L., and Atkinson, P. H. (1983). Correlation of glycosylation forms with position in amino acid sequence. *J. Cell Biol.* **97,** 293–300.

Pongor, S., and Szalay, A. A. (1985). Prediction of homology and divergence in the secondary structure of polypeptides. *Proc. Natl. Acad. Sci. U.S.A.* **82,** 366–370.

Pontier, J. (1970). Traitement des informations chez les êtres vivants: Systèmes à service par essais successifs. *Bull. Math. Biophys.* **32,** 83–148.

Pribnow, D. (1975). Nucleotide sequence of an RNA polymerase binding site at an early T7 promoter. *Proc. Natl. Acad. Sci. U.S.A.* **72,** 784–788.

Proudfoot, N. J., and Brownlee, G. G. (1976). 3′ noncoding region sequences in eukaryotic messenger RNA. *Nature (London)* **263,** 211–214.

Ptashne, M. (1986). Gene regulation by proteins acting nearby and at a distance. *Nature (London)* **322,** 697–701.

Pustell, J., and Kafatos, F. C. (1982). A high speed, high capacity homology matrix: Zooming through SV40 and polyoma. *Nucleic Acids Res.* **10,** 4765–4782.

Quigley, G. J., Gehrke, L., Roth, D. A., and Auron, P. E. (1984). Computer-aided nucleic acid secondary structure modeling incorporating enzymatic digestion data. *Nucleic Acids Res.* **12,** 347–366.

Rapoport, T. A. (1986). Protein translocation across and integration into membranes. *CRC Crit. Rev. Biochem.* **20,** 73–137.

Raupach, R. E. (1984). Computer programs used to aid in the selection of DNA hybridization probes. *Nucleic Acids Res.* **12,** 833–836.

Rawlings, C. D. (1986). "Software Directory for Molecular Biologists." MacMillan, New York.

Reed, R., and Maniatis, T. (1985). Intron sequence involved in lariat formation during pre-mRNA splicing. *Cell (Cambridge, Mass.)* **41,** 95–105.

Rhodes, D., and Klug, A. (1986). An underlying repeat in some transcriptional control sequences corresponding to half a double helical turn of DNA. *Cell (Cambridge, Mass.)* **46,** 123–132.

Richardson, W. D., Roberts, B. L., and Smith, A. E. (1986). Nuclear location signals in polyoma virus large-T. *Cell (Cambridge, Mass.)* **44,** 77–85.

Roberts, R. J. (1985). Restriction and modification enzymes and their recognition sequences. *Nucleic Acids Res.* **13,** r165–r200.

Robson, B., and Suzuki, E. (1976). Conformational properties of amino acid residues in globular proteins. *J. Mol. Biol.* **107,** 327–356.

Rodier, F., Gabarro-Arpa, J., Ehrlich, R., and Reiss, C. (1982). Key for protein coding sequence identification: Computer analysis of codon strategy. *Nucleic Acids Res.* **10,** 391–402.

Rogers, J. (1985). Exon shuffling and intron insertion in serine protease genes. *Nature (London)* **315,** 458–459.

Rogers, S., Wells, R., and Rechsteiner, M. (1986). Amino acid sequences common to rapidly degraded proteins: The PEST hypothesis. *Science* **234,** 364–368.

Rose, G. D., Gierasch, L. M., and Smith, J. A. (1985a). Turns in peptides and proteins. *Adv. Protein Chem.* **37,** 1–109.

Rose, G. D., Geselowitz, A. R., Lesser, G. J., Lee R. H., and Zehfus, M. H. (1985b). Hydrophobicity of amino acid residues in globular proteins. *Science* **229,** 834–838.

Rosenberg, M., and Court, D. (1979). Regulatory sequences involved in the promotion and termination of RNA transcription. *Annu. Rev. Genet.* **13,** 319–353.

Sadler, J. R., Waterman, M. S., and Smith, T. F. (1983). Regulatory Pattern identification in nucleic acid sequences. *Nucleic Acids Res.* **11,** 2221–2231.

Salser, W. (1977). Globin messenger-RNA sequences—Analysis of base-pairing and evolutionary implications. *Cold Spring Harbor Symp. Quant. Biol.* **42,** 985–1002.

Sanger, F., and Tuppy, H. (1951). The amino-acid sequence in the phenylalanyl chain of insulin. *Biochem. J.* **49,** 481–490.

Sanger, F., Nicklen, S., and Coulson, A. R. (1977). DNA sequencing with chain-terminating inhibitors. *Proc. Natl. Acad. Sci. U.S.A.* **74,** 5463–5467.

Sargan, D. R., Gregory, S. P., and Butterworth, P. H. W. (1982). A possible novel interaction between the 3′-end of 18S ribosomal RNA and the 5′-leader sequence of many eukaryotic messenger RNAs. *FEBS Lett.* **147,** 133–136.

Savageau, M. A., Metter, R., and Brockman, W. W. (1983). Statistical significance of partial base-pairing potential: Implications for recombination of SV40 DNA in eukaryotic cells. *Nucleic Acids Res.* **11,** 6559–6570.

Schaller, H., Gray, C., and Herrmann, K. (1975). Nucleotide sequence of an RNA polymerase binding site from the DNA of bacteriophage fd. *Proc. Natl. Acad. Sci. U.S.A.* **72,** 737–741.

Schmidt, G. W., and Mishkind, M. L. (1986). The transport of proteins into chloroplasts. *Annu. Rev. Biochem.* **55,** 879–912.

Schmitz, A., and Galas, D. (1979). The interaction of RNA polymerase and *lac* repressor with the *lac* control region. *Nucleic Acids Res.* **6,** 111–137.

Schneider, T. D., Stormo, G. D., Gold, L., and Ehrenfeucht, A. (1986). Information content of binding sites on nucleotide sequences. *J. Mol. Biol.* **188,** 415–431.

Schulz, G. E., and Schirmer, R. H. (1979). "Principles of Protein Structure." Springer Verlag, Berlin and New York.

Schwyzer, R. (1986). Estimated conformation, orientation, and accumulation of dynorphin A-(1–13) tridecapeptide on the surface of neutral lipid membranes. *Biochemistry* **25,** 4281–4286.

Seeburg, P. H., Nusslein, C., and Schaller, H. (1977). Interaction of RNA polymerase with promoters from bacteriophage fd. *Eur. J. Biochem.* **74,** 107–113.

Sellers, P. H. (1974). On the theory and computation of evolutionary distances. *SIAM J. Appl. Math.* **26,** 787–793.

Serfling, E., Jasin, M., and Schaffner, W. (1985). Enhancers and eukaryotic gene transcription. *Trends Genet.* **1,** 224–230.

Shapiro, B. A., Nussinov, R., Lipkin, L. E., and Maizel, J. V. (1986). A sequence analysis system encompassing rules for DNA helical distortion. *Nucleic Acids Res.* **14,** 75–86.

Sharp, P. M., and Li, W. H. (1987) The codon adaptation index—a measure of directional synonymous codon usage bias and its potential applications. *Nucleic Acids Res.* **15,** 1281–1294.

Sharp, P. M., Tuohy, T. M. F., and Mosurski, K. R. (1986). Codon usage in yeast: Cluster analysis clearly differentiates highly and lowly expressed genes. *Nucleic Acids Res.* **14,** 5125–5143.

Shepherd, J. C. W. (1981a). Method to determine the reading frame of a protein from the purine/pyrimidine genome sequence and its possible evolutionary justification. *Proc. Natl. Acad. Sci. U.S.A.* **78,** 1596–1600.

Shepherd, J. C. W. (1981b). Periodic correlations in DNA sequences and evidence suggesting their evolutionary origin in a comma-less genetic code. *J. Mol. Evol.* **17,** 94–102.

Sheridan, R. P., Dixon, J. S., Venkataraghavan, R., Kuntz, I. D., and Scott, K. P. (1985). Amino acid composition and hydrophobicity of protein domains correlate with their structures. *Biopolymers* **24,** 1995–2023.

Shine, J., and Dalgarno, L. (1974). The 3′-terminal sequence of *Escherichia coli* 16S ribosomal RNA: Complementarity to nonsense triplets and ribosome binding sites. *Proc. Natl. Acad. Sci. U.S.A.* **71,** 1342–1346.

Shpaer, E. G. (1986). Constraints on codon context in *Escherichia coli* genes: Their possible role in modulating the efficiency of translation. *J. Mol. Biol.* **188,** 555–564.

Shuey, D. J., and Parker, C. S. (1986). Bending of promoter DNA on binding of heat shock transcription factor. *Nature (London)* **323,** 459–461.

Shulman, M. J., Steinberg, C. M., and Westmoreland, N. (1981). The coding function of nucleotide sequences can be discerned by statistical analysis. *J. Theor. Biol.* **88,** 409–420.

Sinohara, H., and Maruyama, T. (1973). Evolution of glycoproteins as judged by the frequency of occurrence of the tripeptides Asn-X-Ser and Asn-X-Thr in proteins. *J. Mol. Evol.* **2,** 117–122.

Smeekens, S., Bauerle, C., Hageman, J., Keegstra, K., and Weisbeek, P. (1986). The role of the transit peptide in the routing of precursors toward different chloroplast compartments. *EMBO J.* **5,** 365–375.

Smith, A. E., Kalderon, D., Roberts, B. L., Colledge, W. H., Edge, M., Gillett, P., Markham, A., Paucha, E., and Richardson, W. D. (1985). The nuclear location signal. *Proc. R. Soc. London, Ser. B* **226,** 43–58.

Smith, T. F., and Waterman, M. S. (1981). Identification of common molecular subsequences. *J. Mol. Biol.* **147,** 195–197.

Smith, T. F., Waterman, M. S., and Burks, C. (1985). The statistical distribution of nucleic acid similarities. *Nucleic Acids Res.* **13,** 645–656.

Staden, R. (1980). A computer program to search for tRNA genes. *Nucleic Acids Res.* **8,** 817–825.

Staden, R. (1982a). An interactive program for comparing and aligning nucleic acid and amino acid sequences. *Nucleic Acids Res.* **10,** 2951–2961.

Staden, R. (1982b). Automation of the computer handling of gel reading data produced by the shotgun method of DNA sequencing. *Nucleic Acids Res.* **10,** 4731–4751.

Staden, R. (1984a). Computer methods to locate signals in nucleic acid sequences. *Nucleic Acids Res.* **12,** 505–519.

Staden, R. (1984b). Graphic methods to determine the function of nucleic acid sequences. *Nucleic Acids Res.* **12,** 521–538.

Staden, R. (1984c). Measurements of the effects that coding for a protein has on a DNA sequence and their use for finding genes. *Nucleic Acids Res.* **12,** 551–567.

Staden, R. (1986). The current status and portability of our sequence handling software. *Nucleic Acids Res.* **14,** 217–231.

Staden, R., and McLachlan, A. D. (1982). Codon preference and its use in identifying protein coding regions in long DNA sequences. *Nucleic Acids Res.* **10,** 141–156.

Steitz, J. A., and Jakes, K. (1975). How ribosomes select initiator regions in mRNA: Base pair formation between the 3′-terminus of 16S rRNA and the mRNA during initiation of protein synthesis in *Escherichia coli. Proc. Natl. Acad. Sci. U.S.A.* **72,** 4734–4738.

Sternberg, M. J. E. (1983). The analysis and prediction of protein structure. In "Computing in Biological Science" (Geisow and Barrett, eds.), pp. 143–177. Elsevier, Amsterdam.

Stormo, G. D., Schneider, T. D., and Gold, L. M. (1982a). Characterization of translational initiation sites in *E. coli. Nucleic Acids Res.* **10,** 2971–2996.

Stormo, G. D., Schneider, T. D., Gold, L., and Ehrenfeucht, A. (1982b). Use of the 'perceptron' algorithm to distinguish translational initiation sites in *E. coli. Nucleic Acids Res.* **10,** 2997–3010.

Stormo, G. D., Schneider, T. D., and Gold, L. (1986). Quantitative analysis of the relationship between nucleotide sequence and functional activity. *Nucleic Acids Res.* **14,** 6661–6679.

Stüber, K. (1986). Nucleic acid secondary structure prediction and display. *Nucleic Acids Res.* **14,** 317–326.

Studnicka, G. M., Rahn, G. M., Cummings, I. W., and Salser, W. A. (1978). Computer method for predicting the secondary structure of single-stranded RNA. *Nucleic Acids Res.* **5,** 3365–3387.

Suck, D., and Oefner, C. (1986). Structure of DNase I at 2.0 Å resolution suggests a mechanism for binding to and cutting DNA. *Nature (London)* **321,** 620–625.

Suzuki, H., Kunisawa, T., and Otsuka, J. (1986). Theoretical evaluation of transcriptional pausing effect on the attenuation in *trp* leader sequence. *Biophys. J.* **49,** 425–435.

Sweet, R. M. (1986). Evolutionary similarity among peptide segments is a basis for prediction of protein folding. *Biopolymers* **25,** 1565–1577.

Sweet, R. M., and Eisenberg, D. (1983). Correlation of sequence hydrophobicities measures similarity in three-dimensional protein structure. *J. Mol. Biol.* **171,** 479–488.

Takahashi, K., Vigneron, M., Matthes, H., Wildeman, A., Zenke, M., and Chambon, P. (1986). Requirement of stereospecific alignments for initiation from the simian virus 40 early promoter. *Nature (London)* **319,** 121–126.

Takanami, M., Sugimoto, K., Sugisaki, H., and Okamoto, T. (1976). Sequence of promoter for coat protein of bacteriophage fd. *Nature (London)* **260,** 297–302.

Tanaka, T., Slamon, D. J., and Cline, M. J. (1985). Efficient generation of antibodies to oncoproteins by using synthetic peptide antigens. *Proc. Natl. Acad. Sci. U.S.A.* **82,** 3400–3404.

Taylor, W. R. (1986a). Identification of protein sequence homology by consensus template alignment. *J. Mol. Biol.* **188,** 233–258.

Taylor, W. R. (1986b). The classification of amino acid conservation. *J. Theor. Biol.* **119,** 205–218.

Taylor, W. R., and Thornton, J. M. (1983). Prediction of super-secondary structure in proteins. *Nature (London)* **301,** 540–542.

Taylor, W. R., and Thornton, J. M. (1984). Recognition of super-secondary structure in proteins. *J. Mol. Biol.* **173,** 487–514.

Tessier, L. H., Sondermeyer, P., Fauré, T., Dreyer, D., Benavente, A., Villeval, D., Courtney, M., and Lecocq, J. P. (1984). The influence of mRNA primary and secondary structure on human IFN-γ gene expression in *E. coli. Nucleic Acids Res.* **12,** 7663–7675.

Thornton, J. M., Edwards, M. S., and Barlow, D. J. (1985). Antigenic recognition. *Proc. Eur. 'Oyez' Semin., 2nd.*

Thornton, J. M., Edwards, M. S., Taylor, W. R., and Barlow, D. J. (1986). Location of 'continuous' antigenic determinants in the protruding regions of proteins. *EMBO J.* **5,** 409–413.

Travers, A. A. (1984). Conserved features of coordinately regulated *E. coli* promoters. *Nucleic Acids Res.* **12,** 2605–2618.

Trifonov, E. N., and Bolshoi, G. (1983). Open and closed 5S RNA, the only two universal structures encoded in the nucleotide sequence. *J. Mol. Biol.* **169,** 1–13.

Trifonov, E. N., and Sussman, J. L. (1980). The pitch of chromatin DNA is reflected in its nucleotide sequence. *Proc. Natl. Acad. Sci. U.S.A.* **77,** 3816–3820.

Tsunasawa, S., Stewart, J. W., and Sherman, F. (1985). Amino-terminal processing of mutant forms of yeast Iso-1-cytochrome *c:* The specificity of methionine aminopeptidase and acetyltransferase. *J. Biol. Chem.* **260,** 5382–5391.

Tung, C. S., and Burks, C. (1986). Characterization of the distribution of potential short restriction fragments in nucleic acid sequence databases: Implications for an alternative to chemical synthesis of oligonucleotides. *FEBS Lett.* **205,** 299–302.

Tung, C. S., and Harvey, S. C. (1986). Base sequence, local helix structure, and macroscopic curvature of A-DNA and B-DNA. *J. Biol. Chem.* **261,** 3700–3709.

Tyson, H., and Haley, B. (1985). Alignment of nucleotide or amino acid sequences on microcomputers, using a modification of Sellers' (1974) algorithm which avoids the need for calculation of the complete distance matrix. *Comp. Methods Prog. Biomed.* **21,** 3–10.

Ulanovsky, L., Bodner, M., Trifonov, E. N., and Choder, M. (1986). Curved DNA: Design, synthesis, and circularization. *Proc. Natl. Acad. Sci. U.S.A.* **83,** 862–866.

van den Berg, J. A., and Osinga, M. (1986). A peptide to DNA conversion program. *Nucleic Acids Res.* **14,** 137–140.

Varenne, S., Buc, J., Lloubes, R., and Lazdunski, C. (1984). Translation is a non-uniform process. J. Mol. Biol. **180,** 549–576.

Vass, J. K., and Wilson, R. H. (1984). 'ZSTATS'—A statistical analysis for potential Z-DNA sequences. *Nucleic Acids Res.* **12,** 825–832.

Vogel, H., and Jähnig, F. (1986). Models for the structure of outer-membrane proteins of *Escherichia coli* derived from Raman spectroscopy and prediction methods. *J. Mol. Biol.* **190,** 191–199.

Vondervizst, F., and Simon, I. (1986). A possible way for prediction of domain boundaries in globular proteins from amino acid sequence. *Biochem. Biophys. Res. Comm.* **139,** 11–17.

von Heijne, G. (1981). On the hydrophobic nature of signal sequences. *Eur. J. Biochem.* **116,** 419–422.

von Heijne, G. (1982). A theoretical study of the attenuation control mechanism. *J. Theor. Biol.* **97,** 227–238.

von Heijne, G. (1985). Signal sequences: The limits of variation. *J. Mol. Biol.* **184,** 99–105.

von Heijne, G. (1986a). Towards a comparative anatomy of N-terminal topogenic protein sequences. *J. Mol. Biol.* **189,** 239–242.

von Heijne, G. (1986b). A new method for predicting signal sequence cleavage sites. *Nucleic Acids Res.* **14,** 4683–4690.

von Heijne, G. (1986c). Why mitochondria need a genome. *FEBS Lett.* **198,** 1–4.

von Heijne, G. (1986d). Mitochondrial targeting sequences may form amphiphilic helices. *EMBO J.* **5,** 1335–1342.

von Heijne, G. (1986e). The distribution of positively charged residues in bacterial inner membrane proteins correlates with the trans-membrane topology. *EMBO J.* **5,** 3021–3027.

von Heijne, G., and Blomberg, C. (1979). The concentration dependence of the error frequencies and some related quantities in protein synthesis. *J. Theor. Biol.* **78,** 113–120.

von Heijne, G., and Uhlén, M. (1987). Homology to region X from staphylococcal protein A is not unique to cell surface proteins. *J. Theor. Biol.* (in press).

von Heijne, G., Blomberg, C., and Liljenström, H. (1987). Theoretical modelling of protein synthesis. *J. Theor. Biol.* **125,** 1–15.

von Hippel, P. H., and Berg, O. G. (1986). On the specificity of DNA–protein interactions. *Proc. Natl. Acad. Sci. U.S.A.* **83,** 1608–1612.

Wada, A., and Suyama, A. (1985). Third letters in codons counterbalance the (G + C)-content of their first and second letters. *FEBS Lett.* **188,** 291–294.

Wada, A., and Suyama, A. (1986). Local stability of DNA and RNA secondary structure and its relation to biological functions. *Prog. Biophys. Mol. Biol.* **47,** 113–157.

Wada, A., Yabuki, S., and Husimi, Y. (1980). Fine structure in the thermal denaturation of DNA: High temperature resolution spectrophotometric studies. *CRC Crit. Rev. Biochem.* **9,** 87–144.

Wallace, B. A., Cascio, M., and Mielke, D. L. (1986). Evaluation of methods for the prediction of membrane protein secondary structures. *Proc. Natl. Acad. Sci. U.S.A.* **83,** 9423–9427.

Wartell, R. M., and Benight, A. S. (1985). Thermal denaturation of DNA molecules: A comparison of theory with experiment. *Phys. Rep.* **126,** 68–107.

Waterman, M. S., Smith, T. F., and Katcher, H. L. (1984). Algorithms for restriction map comparisons. *Nucleic Acids Res.* **12,** 237–242.

Welling, G. W., Weijer, W. J., van der Zee, R., and Welling-Wester, S. (1985). Prediction of sequential antigenic regions in proteins. *FEBS Lett.* **188,** 215–218.

Wharton, R. P., and Ptashne, M. (1986). An α-helix determines the DNA-binding specificity of a repressor. *Trends Biochem. Sci.* **11,** 71–73.

Whitelaw, E., and Proudfoot, N. (1986). α-thalassaemia caused by a poly(A) site mutation reveals that transcriptional termination is linked to 3′ end processing in the human $\alpha 2$ globin gene. *EMBO J.* **5,** 2915–2922.

Wickner, W. T., and Lodish, H. F. (1985). Multiple mechanisms of protein insertion into and across membranes. *Science* **230,** 400–407.

Wierenga, R. K., Terpstra, P., and Hol, W. G. J. (1986). Prediction of the occurrence of the ADP-binding β-α-β fold in proteins, using an amino acid sequence fingerprint. *J. Mol. Biol.* **187,** 101–107.

Wierenga, R. K., Swinkels, B., Michels, P. A. M., Osinga, K., Misset, O., Van Beeumen, J., Gibson, W. C., Postma, J. P. M., Borst, P., Opperdoes, F. R., and Hol, W. G. J. (1987). Common elements on the surface of glycolytic enzymes from *Trypanosoma brucei* may serve as topogenic signals for import into glycosomes. *EMBO J.* **6,** 215–221.

Wilbur, W. J., and Lipman, D. J. (1983). Rapid similarity searches of nucleic acid and protein data banks. *Proc. Natl. Acad. Sci. U.S.A.* **80,** 726–730.

Williams, A. L., and Tinoco, I. (1986). A dynamic programming algorithm for finding alternative RNA secondary structures. *Nucleic Acids Res.* **14,** 299–315.

Wolfenden, R. V., Cullis, P. M., and Southgate, C. C. F. (1979). Water, protein folding, and the genetic code. *Science* **206,** 575–577.

Yang, J., Ye, J., and Wallace, D. C. (1984). Computer selection of oligonucleotide probes from amino acid sequences for use in gene library screening. *Nucleic Acids Res.* **12,** 837–843.

Yanofsky, C. (1981). Attenuation in the control of expression of bacterial operons. *Nature (London)* **289,** 751–758.

Yarus, M., and Folley, L. S. (1985). Sense codons are found in specific contexts. *J. Mol. Biol.* **182,** 529–540.

Zenke, M., Grundström, T., Matthes, H., Wintzerith, M., Schatz, C., Wildeman, A., and Chambon, P. (1986). Multiple sequence motifs are involved in SV40 enhancer function. *EMBO J.* **5,** 387–397.

Zhuang, Y., and Weiner, A. M. (1986). A compensatory base change in U1 snRNA suppresses a 5′ Splice Site Mutation. Cell **46,** 827–835.

Zuker, M. (1986). RNA folding prediction: The continued need for interaction between biologists and mathematicians. *Lect. Math. Life Sci.* **17,** 87–123.

Zuker, M., and Sankoff, D. (1984). RNA secondary structures and their prediction. *Bull. Math. Biol.* **46,** 591–621.

Zuker, M., and Stiegler, P. (1981). Optimal computer folding of large RNA sequences using thermodynamics and auxiliary information. *Nucleic Acids Res.* **9,** 133–148.

Index